THE BASICS OF NONMETALS

THE **BASICS** OF NONMETALS

ALLAN B. COBB

ROSEN
PUBLISHING
New York

This edition published in 2014 by:

The Rosen Publishing Group, Inc.
29 East 21st Street
New York, NY 10010

Library of Congress Cataloging-in-Publication Data

Cobb, Allan B.
The basics of nonmetals/Allan B. Cobb.—First edition.
pages cm.—(Core concepts)
Audience: Grades 7 to 12.
Includes bibliographical references and index.
ISBN 978-1-4777-2711-9 (library binding)
1. Nonmetals—Juvenile literature. I. Title.
QD161.C64 2014
546'.7—dc23

2013026878

Manufactured in the United States of America

CPSIA Compliance Information: Batch #W14YA: For further information, contact Rosen Publishing, New York, NY, New York, NY, at 1-800-237-9932.

CONTENTS

CHAPTER ONE

HYDROGEN

Hydrogen is a gaseous element that holds the prime position at the top of the periodic table. It is the element from which all other elements are formed, in a process that starts in the stars.

Hydrogen (H) is the lightest of all the chemical elements. It is also the most common element in the universe. Roughly 75 percent of all matter by mass in the universe is hydrogen. If counting actual atoms, more than 90 percent of all atoms in the universe are hydrogen. On Earth, hydrogen gas is rare in the atmosphere because the atoms are so light that they escape gravity and float into space. Even though it is rare in the atmosphere, hydrogen is the tenth most abundant element on Earth. Water (H_2O) is the most common source for hydrogen. Other sources of hydrogen are methane (CH_4) and hydrocarbons in fossil fuels.

PHYSICAL PROPERTIES OF HYDROGEN

The atomic mass of hydrogen is 1.00794, which is the mass of the proton in its nucleus. Hydrogen is colorless, odorless, and tasteless. At standard temperature and pressure, hydrogen is a gas, H_2. In much of the universe, hydrogen exists as plasma, a high-energy state of matter. In stars, hydrogen undergoes a nuclear fusion reaction because of the

Most of outer space is filled with hydrogen. This galaxy shows vast clouds of hydrogen forming into the pillars that are the birthplace of new blue-white hydrogen stars.

extreme pressure and temperatures that exist. Two hydrogen atoms fuse to form a helium atom. This reaction releases a huge amount of energy as heat and light.

At -32 degrees Fahrenheit (-32°F; 0°C) hydrogen has a density of 0.08988 grams per liter, making it the lightest of all the elements. The melting point of hydrogen is -434.5°F (-259.1°C). The boiling point for hydrogen is -423.2°F (-252.9°C). Because its boiling point is so low, elemental (uncombined) hydrogen always occurs as a gas.

CHEMICAL PROPERTIES OF HYDROGEN

Hydrogen is the first element in the periodic table. It has an atomic number of 1 because it has one proton in its nucleus. A hydrogen atom usually consists of a single proton with a single electron in orbit around it. Protons and electrons are the tiny particles that make up atoms. To create a stable molecule, elemental hydrogen exists as two hydrogen atoms bonded together. This arrangement of two atoms makes hydrogen a diatomic element.

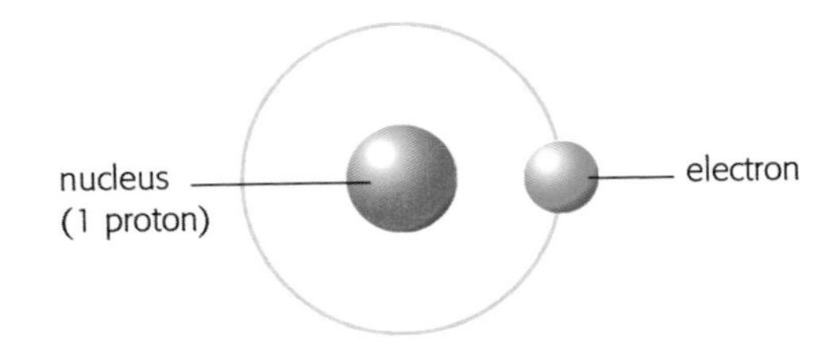

Hydrogen is the simplest of all the elements. It contains only one proton and one electron.

Oil rigs have to get rid of methane gas, which they do by burning. Methane (CH_4) is one of the chief sources of hydrogen gas.

Diatomic means "two atoms" and is a common form for nonmetal elements.

Hydrogen is very reactive. When ignited, it reacts violently with oxygen to form water. Hydrogen also reacts violently with air, halogens, and strong oxidants (reactive oxygen compounds), causing a fire and explosion hazard. These reactions may be enhanced with catalysts (substances that speed up a reaction) such as

KEY DEFINITIONS

- **Atom:** The smallest independent part of an element that still retains its properties.
- **Element:** A substance made up of only one type of atom.

CHEMISTRY IN ACTION

THE HINDENBURG DISASTER

The chemical and physical properties of hydrogen were demonstrated to devastating effect with the *Hindenburg*. The *Hindenburg* was an airship. Built in 1936, it was longer than three Boeing 747s and was the largest airship ever built. To make it lighter than air, the *Hindenburg* was filled with hydrogen. Because hydrogen is such a light element, the lifting power of the vessel was great. The 260,000 cubic yards (200,000 m3) of hydrogen could lift 123.3 tons (112.1 metric tons). It was powered by four diesel motors and had a top speed of 197 miles/hr (135 km/hr). On a journey to the United States, the *Hindenburg* revealed one of the key chemical properties of hydrogen. On May 6, 1937, at the Lakehurst Naval Air Station in New Jersey, the *Hindenburg* erupted into flames. The exact cause of the fire is not known, but it certainly underscores the explosive chemical properties of hydrogen.

The airship *Hindenburg* on fire over a New Jersey naval base. The airship was filled with hydrogen, an extremely flammable gas. It was replaced with completely safe helium gas in later airships.

platinum and nickel. Hydrogen also reacts readily with carbon to form organic chemicals (complex substances that contain large numbers of carbon and hydrogen atoms).

THE IMPORTANCE OF REACTIONS

Because hydrogen is highly reactive and very common, there are many different reactions that involve hydrogen. The combustion of hydrogen in the presence of oxygen yields water, H_2O, and releases energy. Other reactions involving hydrogen gas are often violent because of the considerable energy released. Reactions of hydrogen-containing compounds are usually less violent. Hydrocarbons are burned to provide heat to power engines. Acids react with metals to release hydrogen gas.

UNDERSTANDING HYDROGEN COMPOUNDS

On Earth, water is probably the most important hydrogen compound. Water is vital to life. Living organisms are made up of mostly water. The human body is about two-thirds water. Water is needed to carry dissolved substances around the body. It

Water is the most common compound of hydrogen on Earth. Most living things on the planet need this simple compound to survive.

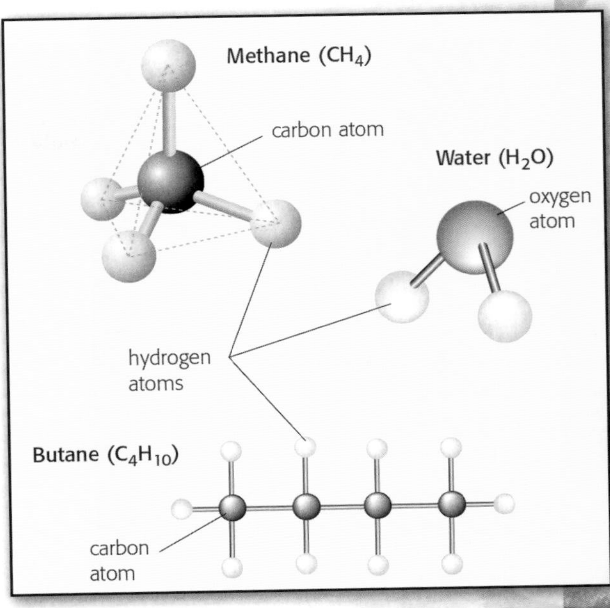

The simplest hydrogen compounds are water, methane, and long-chain hydrocarbons such as butane.

is also needed in many of the biochemical reactions in the body. Plants use water and carbon dioxide to make food and oxygen.

Hydrocarbons are also important hydrogen-containing compounds. There are millions of different hydrocarbons. The simplest of them is methane. Hydrocarbons, other than methane, involve chains of many carbon atoms bonded together. Hydrogen atoms fill all the remaining available bonds on the carbon atoms.

A LOOK AT ISOTOPES

As stated earlier, most hydrogen atoms contain a single proton and electron.

In fact, 99.985 percent of all hydrogen atoms have a single proton and electron. However, there are other isotopes of hydrogen. Isotopes are versions of an element that have the same numbers of protons and electrons but a different number of neutrons. Neutrons are particles with no electrical charge. Hydrogen is the only element where the different isotopes have different names. The most common isotope of hydrogen (that is, a normal hydrogen atom) is called protium. The next isotope of hydrogen has one proton, one neutron, and one electron. This isotope is called deuterium and it accounts for around 0.015 percent of all hydrogen atoms. Deuterium is used to manufacture what is known as "heavy water," D_2O, which is used by the nuclear

Bubbles of hydrogen gas stream off pellets of zinc that have been covered by hydrochloric acid. This is a standard method of producing hydrogen in a laboratory. If a lighted match is placed at the top of the tube, the hydrogen can be heard to go "pop."

industry and in experiments. The third naturally occurring isotope has one proton, two neutrons, and one electron. It is called tritium. Tritium is very rare.

PREPARING HYDROGEN

In the laboratory, the most common method for preparing hydrogen gas is by treating one of many different metals with either an acid or a base. According to the standard definition of an acid, any substance that produces hydrogen ions (H+) is an acid. When an acid is added to a metal, such as zinc, in an aqueous solution, the reaction yields hydrogen gas and a salt:

$$\underset{\text{zinc}}{Zn} + \underset{\text{hydrochloric}}{2HCl} \rightarrow \underset{\text{zinc}}{ZnCl_2} + \underset{\text{hydrogen}}{H_2}$$

According to the standard definition of a base, any substance that produces hydroxide ions, OH−, is a base. When water

CHEMISTRY IN ACTION

DRIVING ON HYDROGEN

Hydrogen-powered cars may become more common in the future as oil begins to run out. However, unlike fossil fuels, elemental hydrogen is not found on Earth. Consequently, any use of hydrogen as a fuel requires the use of energy to make the hydrogen. That does not mean that hydrogen is not a good fuel. Hydrogen is used to power rockets into space. BMW is currently working on a car that runs on hydrogen. One of the main advantages of a hydrogen car is that it would not produce any polluting emissions. The only product from the combustion of hydrogen would be water.

(also a source of hydroxide ions) is added to a highly reactive metal such as sodium (Na), the reaction yields hydrogen gas and a base (sodium hydroxide, NaOH):

$$2Na + 2H_2O \rightarrow 2NaOH + H_2$$

PRODUCTION OF HYDROGEN

Commercial production of hydrogen uses two main methods. The first method is a process that removes hydrogen from hydrocarbons. This commonly used method is called steam reforming of methane. Steam (water vapor) is reacted with methane (CH_4) at a high temperature to yield carbon monoxide (CO) and hydrogen gas. The reaction also takes place at a high pressure so the product, hydrogen gas, is already under pressure and ready to use in industrial processes. The reaction for steam reforming of methane is shown below:

$$CH_4 + H_2O \rightarrow CO + 3H_2$$

The carbon monoxide from the reaction can be used in another reaction with water to create even more hydrogen:

$$CO + H_2O \rightarrow CO_2 + H_2$$

Electrolysis is the other main production method for hydrogen. An electric current can be passed through water to form hydrogen and oxygen gases.

Usually, hydrogen is produced during the chlor-alkali process. In this process, a current is passed through a solution of sodium chloride. The products are chlorine gas, hydrogen gas, and sodium hydroxide.

This apparatus produces hydrogen gas from water using a semiconductor.

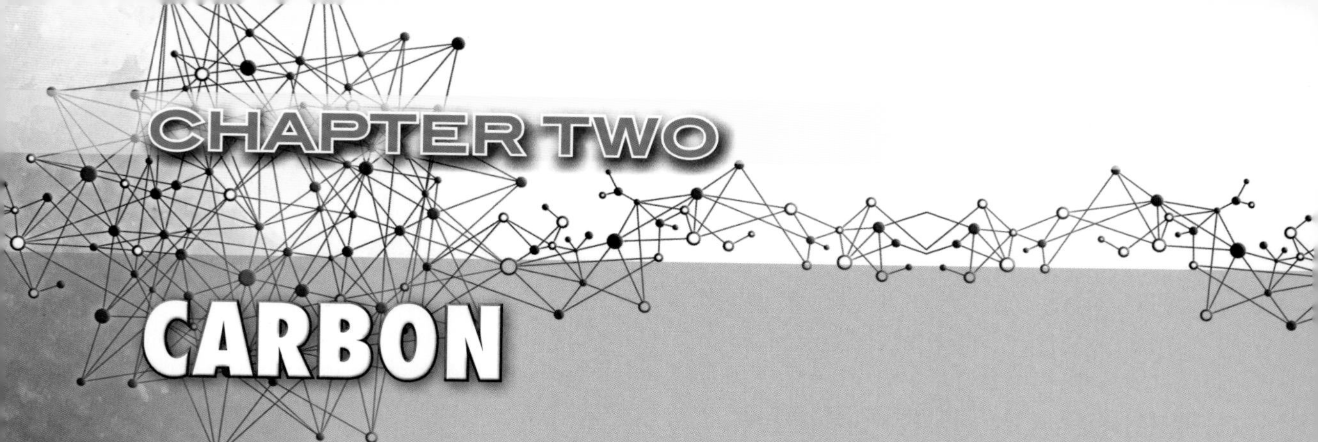

CHAPTER TWO

CARBON

The element carbon is known to form nearly 10 million compounds. Carbon compounds form the basis of all life on Earth. The element occurs in many different varieties, from very hard diamond to very soft graphite.

Carbon is a nonmetal chemical element. In the periodic table, it has the symbol C. Carbon has an atomic number of 6 and an atomic mass of 12.0107. Carbon is the sixth most abundant element in the universe. Carbon is best known because it is the basis of all known life. The study of carbon compounds is usually covered by organic chemistry. In this chapter, the main focus will be on inorganic carbon with only a few references to organic chemistry.

Diamond is one of the best known forms of carbon. It has been treasured as a gemstone for thousands of years, and is also the hardest known substance. Diamonds are cut to show off their brilliant luster.

THE COMPOSITION OF CARBON

Carbon occurs in many different forms, or allotropes. One allotrope of carbon is graphite. Graphite is among the softest known minerals. It is mixed with clay to make pencil lead. Graphite has a density of 2.267 g/cm2. Diamond is a very different allotrope of carbon. Diamond is the hardest naturally occurring mineral. It has a density of 3.513 g/cm3. The very different properties of graphite and diamond have to do with the arrangement of their carbon atoms.

Carbon is almost always found as a solid. Its melting point is 6,381 degrees Fahrenheit (6,381°F; 3,527°C), and its boiling point is 7,281°F (4,027°C). Because these temperatures are so high, carbon does not occur as a liquid or gas on Earth. However, carbon can be found in liquid or gas forms in stars.

THE CHEMICAL PROPERTIES OF CARBON

Carbon occurs in so many different forms because it has many different bonding patterns. Carbon has a half-filled outer shell and is relatively small. This arrangement gives carbon many special properties. To fill its outer shell, carbon needs to form four covalent bonds (it shares electrons with other atoms). These covalent bonds may be single bonds, double bonds, or even triple bonds. Carbon is one of the few elements that form four bonds and the only element that has so many possible bond configurations. Carbon also has the ability to bond to another carbon atom as well as many other elements. This allows long chains of carbon atoms to form. These chains of carbon atoms are the basis of organic chemistry.

ALLOTROPES

The different molecular structures that carbon can take are called allotropes. Carbon has a number of other allotropes apart from graphite and diamond. Amorphous carbon is an allotrope that lacks a crystalline structure. Coal and

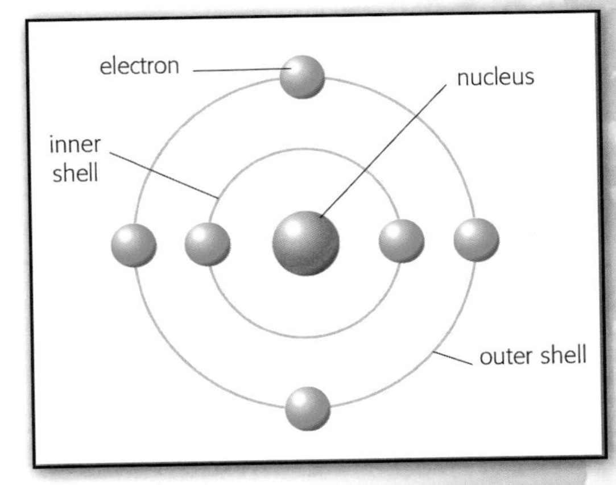

A carbon atom that has not bonded has two electrons in its inner shell and four in its outer, or valence, shell. To fill its outer shell, the atom needs to form four bonds with other atoms. The nucleus contains six protons, which gives carbon an atomic number of 6. The nucleus also contains six neutrons. Adding the masses of the neutrons and protons gives carbon an atomic mass of 12.

When carbon-rich substances burn, as in this apartment fire, the carbon is released as smoke. Smoke, and the sooty residue it leaves behind, is sometimes called amorphous carbon.

soot are sometimes described as amorphous carbon.

The fullerenes are allotropes that were discovered in 1985. Fullerenes were named for the scientist and architect Richard Buckminster Fuller (1895–1983). Fullerenes are molecules made up entirely of carbon atoms in a tiny, hollow sphere or tube shape. The spherical fullerenes are sometimes called buckyballs, and the tube-shaped formations are called nanotubes, or buckytubes. The diameter of a carbon nanotube is about 50,000 times smaller than the diameter of a human hair.

Another fullerene allotrope is called the aggregated diamond nanorod, or ADNR. It is similar in shape to a nanotube but has the same structure as a diamond. The fullerene ADNR is the hardest substance known—it is even harder than diamond.

Another allotrope is carbon nanofoam. This is a low-density cluster-assembly of carbon atoms strung together in a loose three-dimensional web, and it was discovered in 1997. Each cluster is made up of about 4,000 carbon atoms. These are linked together in sheets, like graphite. Carbon nanofoam is only a few times denser than air, and it is a poor electrical conductor.

There are two other allotropes of carbon, but both are rare and have been found only in places where there have been meteor impacts. These allotropes are lonsdaleite and chaoite. Lonsdaleite

A CLOSER LOOK

REMOVING IMPURITIES FROM IRON

Iron ore contains lots of impurities, such as sulfur and oxygen, as well as elemental (uncombined) iron (Fe). People have to remove these impurities before they can use the iron to make steel. The pure iron is removed from the ore through a process called smelting. To remove the impurities, carbon (in the form of coke) is added to the molten iron ore. The carbon acts as a reducing agent (it donates electrons). It bonds with the impurities, which can be removed, and pure elemental iron is left behind.

Molten iron is poured into a wagon at a modern steelworks plant. People have been smelting iron ore for at least 3,000 years.

is a diamond allotrope, and it probably forms where a meteor impact creates great temperature and pressure.

Chaoite is thought to be an allotrope of graphite. It was first found in an impact crater in Bavaria, Germany. Chaoite is slightly harder than graphite and has a slightly different atomic arrangement. However, not all scientists are convinced that chaoite is a true allotrope of carbon.

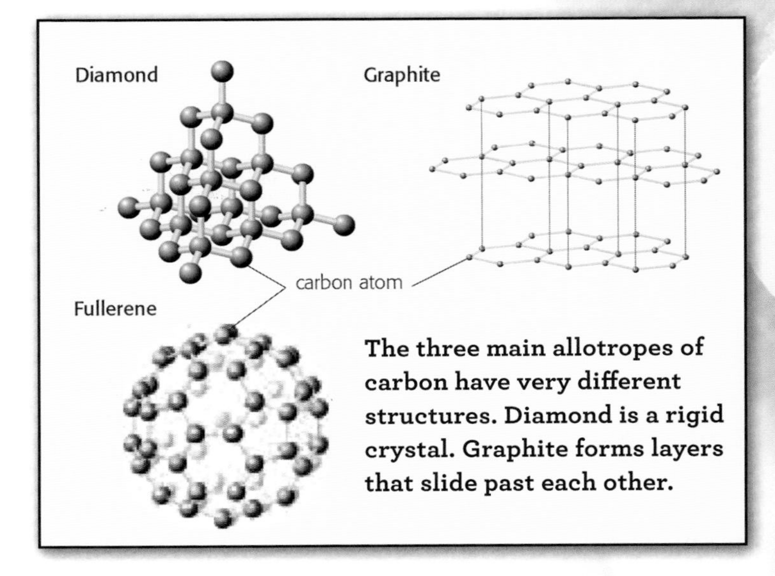

The three main allotropes of carbon have very different structures. Diamond is a rigid crystal. Graphite forms layers that slide past each other.

CHEMISTRY IN ACTION

FAKE DIAMONDS

Artificial diamonds are chemically and structurally identical to natural diamonds. The only difference is that they are made in a laboratory. Exposing carbon to extremely high temperature and pressure makes artificial, or synthetic, diamonds. This process is similar to the way diamonds form naturally. The technique is used to make diamonds that are used in industry as abrasives (substances used for smoothing and polishing). Until 2004, only small shards of synthetic diamonds could be produced. In 2004, a method of making high-quality diamonds as long as 21/2 inches (1 cm) was discovered. These are created from molten graphite at high temperature and pressure.

ISOTOPES OF CARBON

Carbon, like other elements, has a number of different isotopes. Isotopes are forms of an element with the same number of protons in the nucleus but different numbers of neutrons. Thus, isotopes have different mass numbers (the sum of the protons and neutrons). The mass number is added to the name of

The Barringer Crater in Arizona formed when the Canyon Diablo meteorite hit Earth thousands of years ago. Scientists believe the heat and pressure of the impact created tiny crystals of lonsdaleite in the meteorite. Lonsdaleite is a very hard allotrope of carbon, but it is not as hard as diamond.

CHEMISTRY IN ACTION

USING CARBON TO DETERMINE AGE

Carbon-14 is found in the environment, and organisms take it in and incorporate it in their tissues along with carbon-12. The ratio of carbon-12 to carbon-14 in the natural environment is well known, so that is used as the starting point for radiocarbon dating. By looking at the ratio of carbon-12 to carbon-14 in an object, a scientist can determine the age of the object. Because carbon-14 decays (breaks down) at a known rate, the ratio between carbon-12 and carbon-14 changes over time as the carbon-14 decays. The ratio between the two carbon isotopes is then used to determine the age of the material.

This spectrometer can estimate the age of organic materials by working out the proportion of the isotopes of carbon-12 (C-12) and carbon-14 (C-14) present in a sample. Since C-14 decays (breaks down) at a known rate, the machine can find the age of the sample.

the element to identify it. Most carbon occurs in the form, or isotope, carbon-12 or C-12. Carbon-12 is very stable. It has six protons and six neutrons. Carbon-12 accounts for about 98.9 percent of all carbon. Carbon-13, with one additional neutron, is also a stable carbon isotope. It accounts for about 1.1 percent of carbon atoms. Neither of these isotopes undergoes radioactive decay.

There are a total of 15 carbon isotopes, which range from carbon-8 to carbon-22. Apart from carbon-12 and carbon-13, they are rare, and all but one are of limited importance. The only other notable carbon isotope is carbon-14, or radiocarbon. Carbon-14 has 8 neutrons and is not stable. It undergoes radioactive decay into nitrogen-14. The half-life of carbon-14 is 5,730 years.

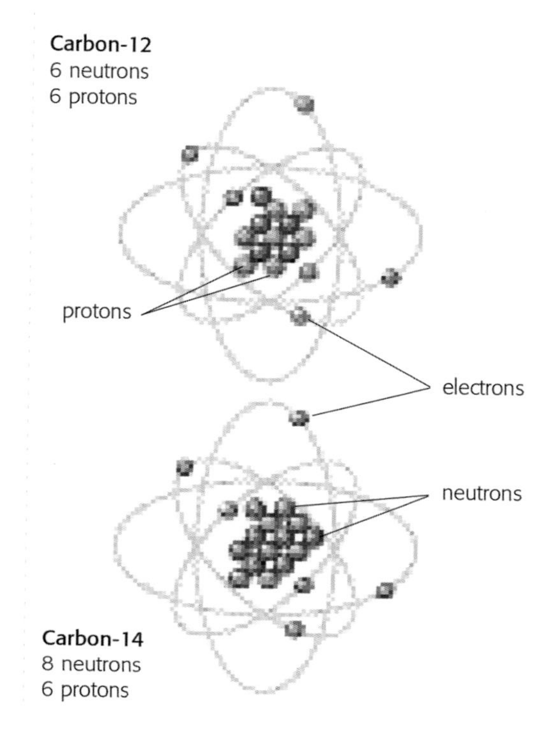

Carbon-12 is the most common isotope of carbon. It is much more stable than another isotope, carbon-14, which gradually decays to form nitrogen-14.

MILLIONS OF CARBON COMPOUNDS

There are more than 10 million different compounds that contain carbon. The vast majority of these are considered organic compounds. They include the alkanes, alkenes, alkynes, amino acids, and fatty acids. These organic compounds are very important as a part of life.

Many different organic compounds are found in petroleum. Petroleum is a complex mixture of various hydrocarbons. Hydrocarbons are chemical compounds that are made up only of carbon and hydrogen, and they have a "backbone" of carbon. Hydrocarbons are extracted from petroleum to make gasoline, diesel, and the petrochemicals used to make plastics. Organic carbon also occurs in coal and natural gas.

Carbon is also found in inorganic compounds. For example, graphite occurs in large quantities in the United States, Russia, Mexico, and India. Diamonds occur in the mineral kimberlite, which is associated with ancient volcanoes. The largest concentrations of diamonds are found in South Africa, Namibia, Congo, Sierra Leone, and Botswana.

Carbon is also found in carbonate rocks such as limestone, dolomite, and marble. Carbonate rocks form as carbonates precipitate out of warm tropical oceans. Limestone contains calcium carbonate ($CaCO_3$), and dolomite is magnesium carbonate ($MgCO_3$). Marble is limestone that has been changed by high pressure and temperature. Carbonate rocks are found in most regions. They represent a huge volume of carbon that is tied up in geological formations.

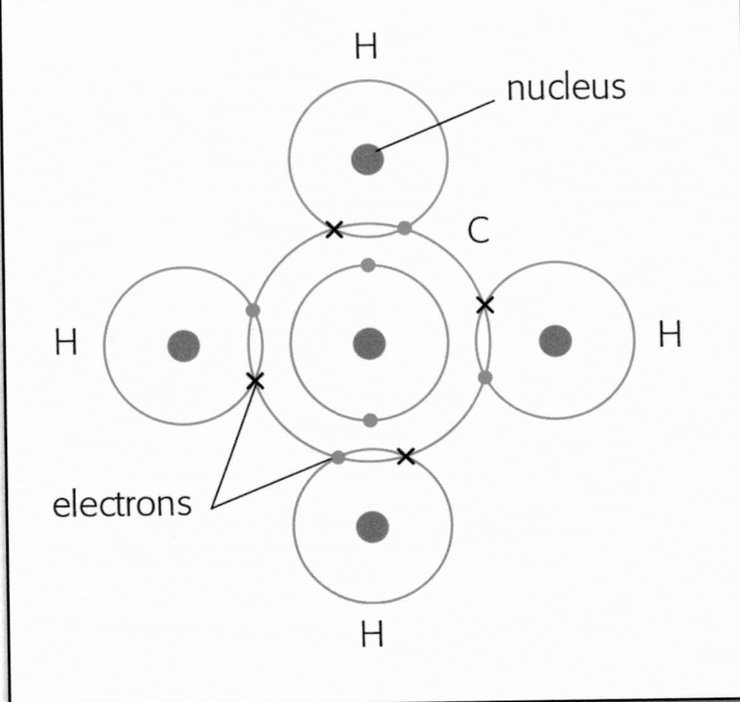

Methane is the simplest hydrocarbon. The valence shell of the carbon (C) atom is filled by the electrons from four hydrogen (H) atoms. More complex hydrocarbons are created when one of the hydrogen atoms is replaced with another carbon.

UNDERSTANDING COVALENT CARBON BONDING

A carbon atom has four electrons in its outer shell. These are called valence electrons. Eight electrons are required to fill the outer shell, so a carbon atom can accept four more electrons. The simplest way to do this is to accept four hydrogen atoms. This creates methane, CH_4, the simplest hydrocarbon. One of the hydrogen atoms can be replaced with a carbon atom, forming a chain. This is how hydrocarbon chains are formed.

Because the electrons are shared between carbon and other elements, these are called covalent bonds.

Covalent bonding frequently occurs in elements with similar electronegativities. The electronegativity of an element is the measure of its ability to attract other electrons. Nonmetals do not give up electrons easily, so sharing electrons is the best way to fill their outer shell. As mentioned earlier, carbon can have single, double, or triple bonds. Carbon

A digger transfers coal onto a truck in an open-cast mine.

typically bonds with elements such as other carbon atoms, hydrogen, nitrogen, sulfur, oxygen, and chlorine.

Because carbon has similar electronegativity to several other elements, it is able to be a part of millions of different compounds. By linking into chains and even circles, carbon atoms are capable of producing a wide array of different compounds. The ability of carbon to make long chains is called catenation. The carbon–carbon bonds that form are fairly strong and unusually stable.

A CLOSER LOOK AT HYDROCARBONS

The simplest of the carbon–carbon compounds are the hydrocarbons. An alkane is a hydrocarbon that has only single bonds between the carbon atoms. Propane (C_3H_8) and butane (C_4H_{10}) are common examples of alkanes. When a hydrocarbon has only single bonds between carbons, it has the maximum possible number of hydrogen atoms. Therefore, an alkane is called a saturated hydrocarbon.

Alkenes have at least one double bond between carbon atoms. Alkynes have at least one triple bond between carbon atoms. Since alkenes and alkynes do not have the maximum number of hydrogen atoms, they are called unsaturated hydrocarbons. Alkenes and alkynes can go through a process called hydrogenation. This breaks the double or triple bonds and adds hydrogen atoms, thus changing the alkenes and alkynes into alkanes. Hydrocarbons can also be linked in such a way that they form a circle. These are called cyclic hydrocarbons.

A special type of cyclic hydrocarbon is an aromatic hydrocarbon. An aromatic hydrocarbon has a molecular structure of six carbon atoms in a ring. Three double bonds occur in the ring, so there are six hydrogen atoms attached. The simplest aromatic hydrocarbon is benzene (C_6H_6), which has only one ring. Some aromatic hydrocarbons have more than one ring.

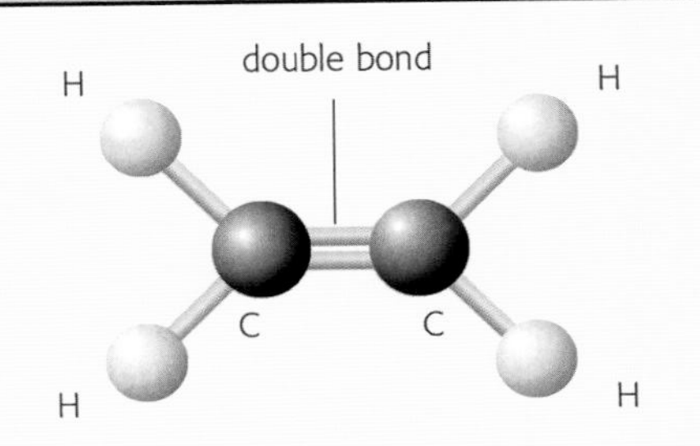

Ethene is the simplest alkene. Two carbon atoms (C) are double bonded, and each is bonded with two hydrogen atoms (H).

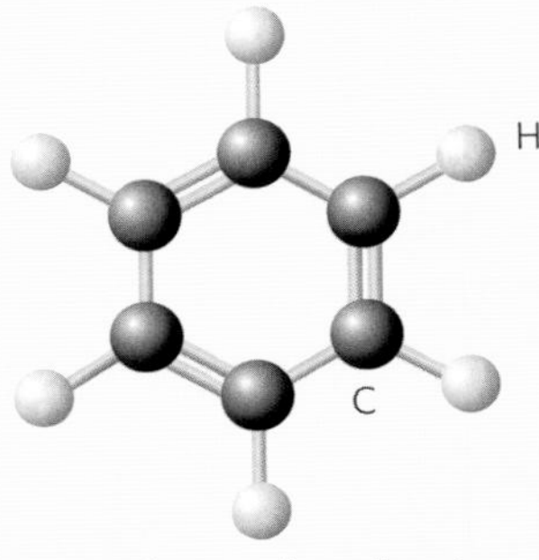

Benzene is used to make plastics, rubbers, detergents, and dyes. It is the simplest of the aromatic hydrocarbons. Six carbon atoms (C) are bonded to six hydrogen atoms (H).

OTHER CARBON COMPOUNDS

Not all carbon compounds fall within the realm of organic chemistry. These other carbon compounds do not come from living or organic sources, so they are called inorganic carbon compounds. Another definition used to distinguish organic and inorganic compounds is whether or not carbon is bonded to hydrogen. Neither of these definitions is foolproof. In general, all the oxides, inorganic salts, cyanides, cyanates, isocyanates, carbonates, and carbides are considered inorganic carbon compounds.

The oxides of carbon are very important. The most common oxide is carbon dioxide, CO_2. Carbon dioxide is a compound that links organic and inorganic carbon compounds. Carbon dioxide is breathed out by humans as well as by plants and animals as they break down food molecules. Plants take in carbon dioxide and use the sun's energy to convert it into food. This is part of the carbon cycle, which is discussed in detail later in this chapter.

Carbon dioxide is also a product of combustion reactions. When a hydrocarbon burns in the presence of oxygen, it produces carbon dioxide and water. If there is insufficient oxygen present during combustion, another oxide of carbon forms—carbon monoxide, CO. Carbon monoxide is a deadly gas. In animals, it binds to hemoglobin (the protein that carries oxygen in the bloodstream) in the blood and prevents oxygen from bonding. If the concentration of carbon monoxide is great enough, this can lead to death.

When carbon dioxide dissolves in water, it forms carbonic acid (H_2CO_3):

$$CO_2 + H_2O \rightarrow H_2CO_3$$

This weak acid is strong enough to dissolve limestone (calcium carbonate, $CaCO_3$). When limestone is dissolved underground, caves form. Calcium carbonate reacts with the carbonic acid in water. Because calcium carbonate and

KEY DEFINITIONS

- **Allotropes:** Different physical forms of a single element.
- **Covalent bonds:** Bonds in which two or more atoms share electrons.
- **Isotopes:** Forms of an element, each having a different mass.

The Taj Mahal in India is made of marble. Marble is a form of limestone, which is a carbon-containing rock.

carbonic acid are both carbon-containing compounds, this sets up an equilibrium reaction (one that works both ways) between the dissolved carbon dioxide, the carbonate ion (CO_{32}–), and the bicarbonate ion (HCO_3–):

$$CaCO_3 + CO_2 + H_2O \leftrightarrow Ca(HCO_3)_2$$

In limestone caves, this equilibrium is important because it determines whether limestone will be dissolved or deposited by precipitation (becoming insoluble again). If the water is too acidic, limestone is dissolved. If too many bicarbonate ions are present, calcium carbonate will be deposited. This deposition is responsible for the growth of stalactites and stalagmites. This equilibrium is also important in seawater as it controls the precipitation of calcium carbonate, which will later become limestone.

CARBON'S INDUSTRIAL USES

Carbon is an important element in industry. Carbon, in the form of coke, is used to

Stalactites and stalagmites in a limestone cave. The spikes growing down from the roof are stalactites, and those growing up from the floor are stalagmites. These features form when calcium carbonate in the rock is dissolved by water. As the water drips, some of the calcium carbonate is precipitated (comes out of the solution as a solid) and builds up over time to form long columns of rock.

remove impurities from smelted iron. At high temperatures, carbon combines with iron to make steel. The amount of carbon used determines the type of steel produced. Steel with a content of about 1.5 percent carbon is used for tools and sheet steel. Steel with about 1 percent carbon is used to make automotive and aircraft parts. High-strength steel used in structural supports has about 0.25 percent carbon.

Both natural and synthetic diamonds are used to grind and drill other materials (they are abrasives). Diamonds are found in grinding disks, drill points, and abrasive powders. Amorphous carbon is another allotrope of carbon that has industrial uses. Amorphous carbon is usually produced by the incomplete combustion of methane. This amorphous carbon is called carbon black, and it is widely used as a filler and reinforcing agent in rubber.

Another use of carbon is as activated charcoal. Activated charcoal is charcoal that has been treated with oxygen to open up spaces between the carbon atoms. Activated charcoal is used to adsorb, or take up, odors and other impurities from liquids and gases.

To be adsorbed, an impurity has to be chemically attracted to the carbon atoms. Activated charcoal is very effective because it has a very large surface area. One gram (0.035 ounces) of activated charcoal may have a surface area of 360–2,400 square yards (300–2,000 m2).

Carbon is also important in the plastics industry. The petrochemicals used to make plastics come from petroleum. The petrochemicals are mainly hydrocarbons extracted from the petroleum. The hydrocarbons are linked together through a process called polymerization to make the different plastics. Plastics are very useful because they can be formed or molded into many different shapes. This property explains why there are so many different plastic products in regular daily use.

A LOOK AT THE CARBON CYCLE

Carbon is a compound that passes through a cycle in the environment called a biogeochemical cycle. All the

A worker supervises the manufacture of plastic products.

CHEMISTRY IN ACTION

THE USES OF CARBON FIBER

Carbon fiber generally refers to the carbon thread or fabric woven from carbon filaments. Carbon filaments are long strands of carbon that form when a plastic fiber is heated. Each fine carbon-fiber thread is stronger than steel. When these fibers are embedded in plastic or epoxy, the resulting material is incredibly strong but very light in weight. Carbon fibers are used to make many products that require great strength. These include sporting equipment, auto parts, tools, boats, and even strings for musical instruments.

The frame of this bicycle is very strong and light. It is made of carbon fiber.

carbon in animals and plants comes from the environment. The source of carbon for living organisms is the atmosphere. Although the atmosphere contains only about 0.038 percent carbon dioxide, this small amount is very important.

Plants convert carbon dioxide from the atmosphere into organic carbon molecules in a process called photosynthesis. During this process the plant combines carbon dioxide and water using energy from the sun to make glucose. The plants then

Dead plants and animals will build up in layers on the bottom of this swamp. The carbon that they contain may eventually form oil and coal in millions of years' time.

convert glucose into other compounds, which they store for later use. When animals eat plants, they use the stored food for energy and to build tissues. So carbon is transferred from plants to animals.

Carbon in the form of carbon dioxide is important for plants to make food. Plants and animals break down carbon compounds in a process called cellular respiration. This process releases both stored energy and carbon dioxide and returns carbon dioxide to the atmosphere. Trees tie up carbon in their wood. When trees burn, the carbon is released as carbon dioxide.

When plants and animals die, they decompose, or break down. When they decompose, carbon is released directly back into the atmosphere. Sometimes, the remains of plants or animals are buried quickly or decompose in swamps where there is little oxygen. When this happens, the carbon is tied up and does not return immediately to the atmosphere. The carbon may then be locked away for thousands or even millions of years. Geological processes, heat, and pressure may turn this trapped carbon into oil, coal, or natural gas. Because these forms of carbon are locked away for

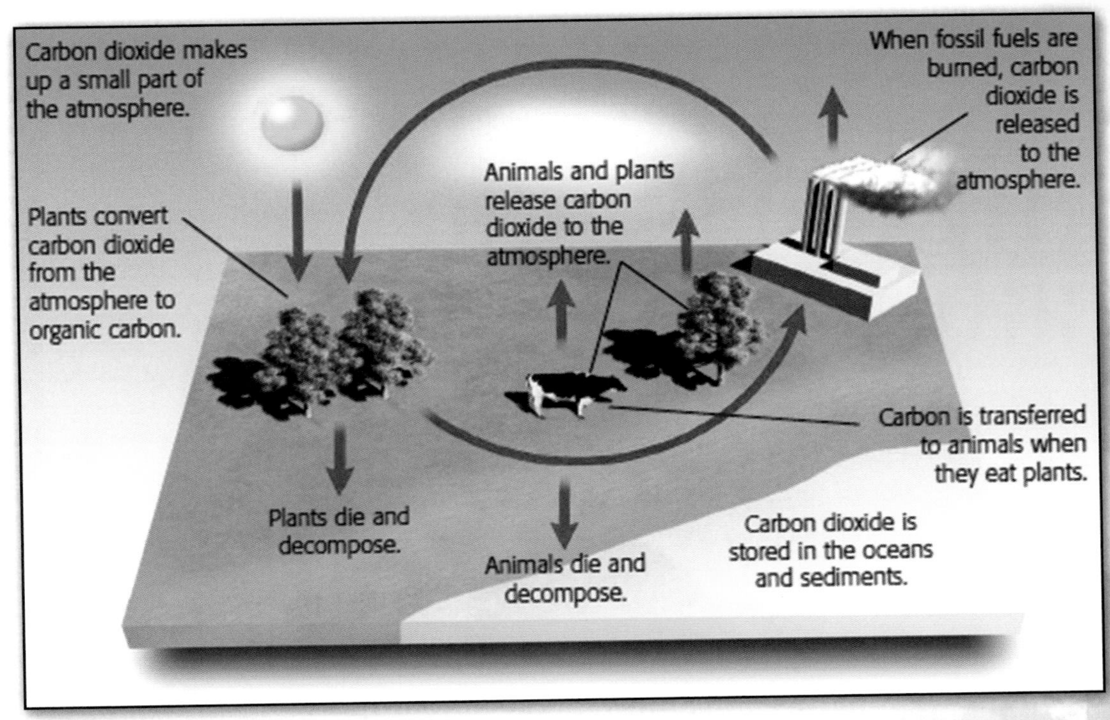

long periods of time, they are called fossil fuels. When people burn fossil fuels for energy, the carbon is released back into the atmosphere as carbon dioxide.

The formation of fossil fuels is not the only way carbon is tied up. Large amounts of carbon dioxide are dissolved in the oceans. Some organisms that live in the oceans convert this carbon dioxide into calcium carbonate for their shells. When the organisms die, their shells fall to the bottom of the ocean. In some oceans, chemical processes cause carbon dioxide dissolved in ocean water to form calcium carbonate. This settles on the ocean floor. Given enough time, the shells and calcium carbonate build up and turn into limestone. This process may tie up carbon for millions of years. Later, erosion slowly releases this carbon back into the environment.

The carbon cycle can be thought of as four connected reservoirs of carbon. They are the atmosphere, the biosphere (land plants and animals), the oceans, and sediments (materials deposited by water, such as mud or sand).

The carbon cycle is a delicate balance between carbon in the atmosphere and carbon that is tied up in organisms and the environment. The balance is altered when fossil fuels are burned, but scientists do not know exactly how the mechanism works.

CHAPTER THREE

NITROGEN AND PHOSPHORUS

Nitrogen is the most common gas in Earth's atmosphere and is an essential element for living things. Phosphorus is also essential for organisms and occurs in three main forms: white phosphorus, red phosphorus, and black phosphorus.

Nitrogen and phosphorus are found in Group 15 of the periodic table. The element nitrogen is represented by the symbol N and has an atomic number of 7. The atomic number is determined by the number of protons in an atom's nucleus. Nitrogen usually occurs as an odorless, colorless, and mostly inert (unreactive)

Earth's atmosphere is nearly 80 percent nitrogen. Nitrogen in air exists as molecules of two bonded nitrogen atoms, or diatomic nitrogen (N_2).

gas. Elemental nitrogen is a gas consisting of molecules of nitrogen formed from two bonded nitrogen atoms. Molecules of this sort are called diatomic. Diatomic nitrogen has the formula N_2. Nitrogen gas makes up 78.084 percent of the atmosphere by volume and is the fifth most common element in the universe. Nitrogen is also an important element to living organisms. Nitrogen is found in all living tissues. Common nitrogen-containing compounds include ammonia (NH_3), nitric acid (HNO_3), cyanides, and amino acids.

The element phosphorus is represented by the symbol P. Phosphorus has an atomic number of 15. Phosphorus is commonly found in inorganic (not containing carbon) phosphate rocks and in living organisms. Phosphorus is highly reactive and never found in nature in its elemental form. Phosphorus emits a faint glow when exposed to oxygen. The name *phosphorus* comes from a Greek word meaning "bearer of light." Phosphorus is widely used in making fertilizers, explosives, fireworks, nerve agents (chemical weapons), pesticides, detergents, and toothpaste.

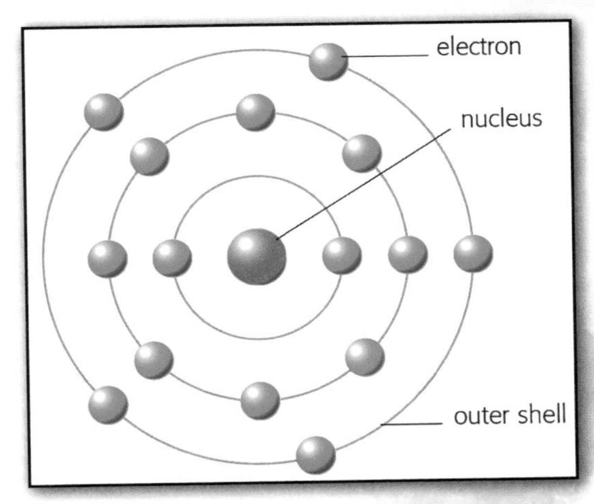

Each phosphorus atom has 15 electrons surrounding its nucleus.

The test tube contains white phosphorus submerged in water. White phosphorus ignites spontaneously in air, so it must be stored in water. Below the test tube is red phosphorus. This form of phosphorus is less volatile.

DIFFERENT FORMS OF PHOSPHORUS

Phosphorus exists in three different forms, or allotropes. These are white phosphorus, red phosphorus, and black phosphorus. White and red phosphorus are the most common. They both consist of four phosphorus atoms arranged in a tetrahedron (pyramid with a triangular base). In white phosphorus, the tetrahedrons form a regular repeated structure, or crystal. White phosphorus is a poisonous waxy solid with an odor like garlic. White phosphorus is very reactive in air and ignites easily. It is therefore commonly stored underwater. In red phosphorus, tetrahedrons are linked in chains. Red phosphorus is less dangerous than the white form and does not ignite spontaneously. Black phosphorus can be obtained by subjecting white phosphorus to high temperatures. Black phosphorus is less reactive than white phosphorus or red phosphorus. It consists of a network of phosphorus atoms with each atom

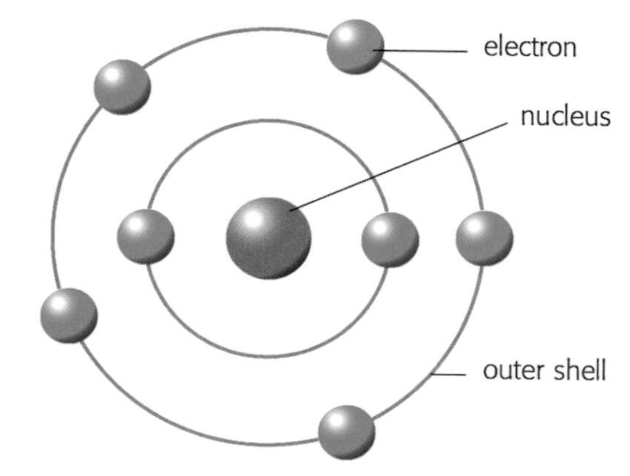

Each nitrogen atom has seven electrons surrounding its nucleus.

A CLOSER LOOK

NITROGEN FIXATION

Some bacteria have developed a mechanism for removing nitrogen gas from air and building it into protein with a process called nitrogen fixation. Many of the bacteria benefit from a close relationship with plants, such as clover *(right)*, which need nitrogen. Therefore the plants also benefit. This form of relationship is termed symbiotic. The bacteria live in clusters called nodes on the roots of some plants such as beans, alfalfa, and peanuts. These plants actually enrich the soil with excess soluble nitrogen (nitrogen dissolved in water). The process of nitrogen fixation by the bacteria is described in simplified form by this chemical equation:

$$3CH_2O + 2N_2 + 3H_2O + 4H^+ \rightarrow 3CO_2 + 4NH_4^+$$

in which formaldehyde (CH2O) reacts with nitrogen, water, and hydrogen ions (H^+) to form carbon dioxide (CO_2) and ammonium ions (NH_4^+).

PROFILE

DANIEL RUTHERFORD (1749–1819)

Scottish chemist Daniel Rutherford discovered nitrogen. His teacher, Joseph Black, was studying carbon dioxide and discovered that if a candle was burned in an upside-down glass in a bowl of water, the water would gradually rise in the glass and the candle would eventually go out. He passed this experiment on to Rutherford. Rutherford kept a mouse in a confined quantity of air until it died. He removed other gases from the air by burning phosphorus and then filtering out carbon dioxide. In the remaining gas, nitrogen, it was not possible for a mouse to survive, or for phosphorus to combust, or a flame to burn. Because of this characteristic, Rutherford named the substance noxious air, or phlogisticated (burned) air.

Daniel Rutherford discovered nitrogen by a process of eliminating the other gases from air.

attached to three others. Black phosphorus has no significant commercial uses.

GROUP 15

Nitrogen and phosphorus are part of the nitrogen group, Group 15. The other members of the group are arsenic, antimony, and bismuth. The elements of Group 15 become increasingly metallic as they increase in atomic number down the group. This trend is reflected both in their structures and in their chemical properties. Nitrogen is unreactive.

Berthold Schwart was a 14th-century German monk. He is credited with being the first European to discover gunpowder.

The only element that reacts with nitrogen at room temperature is lithium, forming lithium nitride, Li_3N. Magnesium also reacts directly with nitrogen, but only when ignited. Phosphorus is more reactive than nitrogen. It reacts with various metals to form phosphides, with sulfur to form sulfides, with halogens to form halides, and forms oxides with oxygen when ignited in air. Phosphorus also reacts with both alkalis and concentrated nitric acid.

KEY DEFINITIONS

- **Nitrogen fixation:** Processes that combine atmospheric nitrogen with other elements in a form that can be absorbed by plants. Nitrogen is fixed chiefly by leguminous plants, such as beans, alfalfa, and clover, but is also fixed by lightning.

NITROGEN'S DISCOVERY

Nitrogen compounds have been known for much longer than nitrogen has been known as an element. Saltpeter—sodium or potassium nitrate—was used in making gunpowder. Gunpowder was first made in China in the 9th century. Later, saltpeter was used as a fertilizer. Nitrogen compounds were well known to the alchemists of the Middle Ages. Nitric acid, called aqua fortis, was first synthesized around 800 CE in the Middle East. Alchemists soon discovered that nitric acid could be mixed with hydrochloric acid to create aqua regia, an acid that would dissolve gold.

Elemental nitrogen was discovered by Scottish chemist Daniel Rutherford (1749–1819) in 1772. Other chemists continued his work, and in 1776 French

Lightning produces high temperatures and pressures that cause nitrogen in the atmosphere to combine with oxygen, producing nitrogen oxide (NO) and nitrogen dioxide (NO_2). The nitrogen dioxide may then dissolve in rainwater producing nitric acid (HNO_3), which plants can then absorb.

This smog over Prague in the Czech Republic is caused by vehicles producing nitrogen oxides. These compounds of nitrogen react with sunlight to produce smog.

chemist Antoine Lavoisier (1743–1794) suggested that this gas was an element.

IN THE AIR

Nitrogen is the largest single component of air. The mass of the nitrogen in the atmosphere is estimated at 4,000 trillion tons (3,628 trillion metric tons). Some bacteria are capable of fixing nitrogen gas into a soluble form usable by plants. Lightning also has the ability to convert nitrogen into a soluble form.

In the atmosphere, nitrogen is four times as abundant as oxygen. However, oxygen is about 10,000 times more abundant than nitrogen on Earth. Oxygen is a major component of the solid earth. Because nitrogen does not form a stable crystal lattice (regular repeated structure), it is rarely incorporated into rocks and minerals. This is one reason why nitrogen has a higher concentration than oxygen in the atmosphere. The other primary reason is that, unlike oxygen, nitrogen is very stable in the atmosphere and does not take part in many chemical reactions. As a result, nitrogen has built up in the atmosphere to a much greater extent than oxygen.

NITROGEN OXIDES

Nitrogen also forms several different oxides—molecules with oxygen. Dinitrogen oxide, N_2O, is an anesthetic called laughing gas. Nitrogen monoxide, NO, and nitrogen dioxide, NO_2, both form when combustion of hydrocarbons in air takes place under high pressure. These nitrogen oxides are produced in internal combustion engines. In the atmosphere, these nitrogen oxides can form smog, a type of atmospheric haze produced by the action of sunlight with pollutants. Two other compounds of nitrogen and oxygen, dinitrogen trioxide, N_2O_3, and dinitrogen pentoxide, N_2O_5, are unstable and explosive.

Two important acids also contain nitrogen—nitrous acid, HNO_2, and nitric acid, HNO_3. These acids can be used to make chemicals called nitrites and nitrates respectively. Nitric acid is a strong acid and has many industrial uses.

Ammonia, NH_3, is probably the most important of all the nitrogen compounds. Ammonia is used as a nutrient by plants, and it bonds with organic carbon molecules to form chemicals called amines. These amines may be assembled into amino acids, which are compounds vital to living organisms. Nitrogen is also part of other classes of organic (carbon-containing) molecules such as the amides, nitro groups, imines, and enamines.

Cheese is high in proteins. Proteins are important biological molecules that contain nitrogen.

NITROGEN AND PROTEINS

Amino acids are nitrogen-containing compounds that serve as the building blocks of proteins. Proteins are important organic compounds made of nitrogen, carbon, hydrogen, and oxygen, and most also contain sulfur. Proteins are essential in the structure and function of all living organisms. Proteins perform many different roles. Some proteins are structural, such as those that provide the structural support (the cytoskeleton) in cells. Other proteins are used to store and transport

CHEMISTRY IN ACTION

NITROGEN OXIDES AND ACID RAIN

The nitrogen oxides that form in internal combustion engines do more than just cause smog. The nitrogen oxides react with water vapor in the atmosphere to create nitrous and nitric acids. In the atmosphere, the gaseous acids dissolve in rainwater and create acid precipitation, commonly called acid rain. Acid rain can accelerate the weathering rate for rocks, including those used in buildings and sculptures.

The decorative stonework on this building has been damaged by acid rain. The acid has reacted with minerals in the stone.

materials in cells. Still other proteins, called enzymes, speed up reactions in organisms. Proteins are also an essential part of diet because they serve as an important nitrogen source for organisms.

Organisms synthesize many amino acids, but others must be obtained from food. More than 100 amino acids have been discovered. Both plants and animals produce amino acids. Amino acids have also been detected in meteorites and comets.

THE PRODUCTION OF NITROGEN

Nitrogen is widely used in industry. Industrial production of nitrogen is performed by one of three processes. These are the pressure swing adsorption method, the diffusion separation process, and cryogenic distillation.

CHEMISTRY IN ACTION

THE USES OF LIQUID NITROGEN

The study of very low temperatures is called cryogenics. Cryogenics is carried out using liquid gases, and liquid nitrogen is the most commonly used cryogenic substance. Liquid nitrogen has many useful applications. In medicine, it is used for freezing areas of skin to treat skin cancers and remove warts. It is also used for freezing human blood, sperm, and embryos so that they can be used at a later date. It is used by food industries for rapid freezing. When the food thaws, bacteria do not flourish because the oxygen in the food has been replaced with nitrogen. Liquid nitrogen can be pumped down oil wells to increase the pressure at the bottom of the wells. This forces the oil to the surface. Steel can be hardened using liquid nitrogen. The steel is plunged into liquid nitrogen, which removes impurities in the structure of the steel, making it less brittle. More controversially, liquid nitrogen is sometimes used to freeze the bodies of people who have died. This is done in the hope that the freezing process will preserve the body so that it can be revived at some time in the future.

Liquid nitrogen being poured into a flask. The vapor seen around the flask is made of water droplets that have condensed from the air because of the extremely low temperature.

A CLOSER LOOK

ANHYDROUS AMMONIA

Anhydrous ammonia (ammonia without water added) is used as fertilizer throughout the world. Ammonia is produced by the Haber process. Each year, more than 500 million tons of fertilizer is produced using this method. The industrial production of ammonia uses about 1 percent of the total world's energy and produces fertilizer for about 40 percent of the world's population.

Anhydrous ammonia is injected into soil, where it is absorbed into the soil moisture.

The pressure swing adsorption method uses adsorbents, which are solid materials that attract molecules to their surface. In the pressure swing adsorption method, compressed air is forced through reaction vessels that contain different adsorbents. Each adsorbent has an affinity (attraction) for certain chemicals in air such as oxygen, carbon dioxide, and argon. These gases are removed, leaving only nitrogen. Diffusion separation is a similar method. Air under pressure is pumped into a reaction vessel and membranes (sheetlike tissues) allow only certain gases to pass through. This method filters out the unwanted gases, leaving only nitrogen behind. Both of these methods are commonly used, but the nitrogen produced still has some impurities.

Cryogenic distillation produces ultrapure nitrogen. This process requires considerable energy, but it results in liquid nitrogen. Air is cooled, and all the water vapor and carbon dioxide are removed. The remaining air is then compressed and chilled through a number of steps until it is liquefied. The different gases are distilled out of the liquid air. In distillation, the liquid air is gradually heated, and each of the gases that make up air boil off at specific temperatures. Each gas can then be removed individually. This process yields liquid nitrogen as well as liquid oxygen and liquid argon.

KEY DEFINITIONS

- **Distillation:** The process of boiling a liquid and condensing the vapor to purify it. Also used as a method for separating a liquid mixture into its component pure substances.

THE PRODUCTION OF NITRIC ACID

In the laboratory, nitric acid (HNO_3) can be made from copper nitrate ($Cu(NO_3)_2$) or potassium nitrate (KNO_3) in 96 percent concentrated sulfuric acid (H_2SO_4). Nitric acid is then distilled out of the solution.

The Ostwald process is the industrial method used to produce nitric acid. The process was patented by Latvian-born German chemist Wilhelm Ostwald (1853–1932) in 1902 and is still used.

This process starts with ammonia (NH_3). The ammonia is oxidized (combined with oxygen), which produces nitrogen oxide (NO) and water (H_2O). Platinum and rhodium are used as a catalyst to this reaction. A catalyst is a substance that speeds up a reaction without itself undergoing any chemical change. The reaction takes the following form:

$$4NH_3 + 5O_2 \rightarrow 4NO + 6H_2O$$

The nitrogen oxide is oxidized to form nitrogen dioxide (NO_2):

$$2NO + O_2 \rightarrow 2NO_2$$

The nitrogen dioxide is absorbed by water to yield dilute nitric acid (HNO_3) and nitrogen oxide:

$$3NO_2 + H_2O \rightarrow 2HNO_3 + NO$$

The nitrogen oxide is then recycled, oxidized, and more nitric acid is produced:

$$4NO_2 + O_2 + 2H_2O \rightarrow 4HNO_3$$

The nitric acid is then concentrated to the desired strength by distillation. This process is very efficient and has an overall yield of about 96 percent. The yield is how much of the initial reactants are converted into the final product.

A LOOK AT THE HABER PROCESS

The Haber process is the reaction of nitrogen and hydrogen to produce ammonia. German chemist Fritz Haber (1868–1934) patented this process in 1908. The Haber process is used to make anhydrous (without water) ammonia, ammonium nitrate, and urea for the fertilizer industry. The reaction for the process appears to be quite simple. The arrow pointing in both directions shows that the reaction can work in both directions until an equilibrium state is reached:

$$N_2 + 3H_2 \leftrightarrow 2NH_3$$

However, this reaction takes place over an iron catalyst at 200 times atmospheric pressure and temperatures between 840 and 930°F (450–500°C). The yield of this reaction is only 10 to 20 percent. The reaction reaches an equilibrium state, but as long as product is removed and reactants are added, the reaction continues. When the high-pressure ammonia gas is removed, it becomes a liquid upon cooling.

DISCOVERING PHOSPHORUS

German alchemist Hennig Brand (c. 1630–1710) discovered phosphorus in 1669. He extracted phosphorus from urine. An interesting property of phosphorus was noted immediately and was responsible for continued experiments. It was discovered that phosphorus would glow.

Scientists soon found that if placed in a sealed jar, phosphorus would glow for a period of time and then stop. Irish chemist Robert Boyle (1627–1691) observed that the glow stopped when all the oxygen in the bottle had been consumed. It was soon found that the glow only occurred when there were certain amounts of oxygen. Too much or too little oxygen and the phosphorus would not glow. The mechanism for the glow was not explained until 1974.

Physicists R. J. Van Zee and A. U. Khan discovered the source of the glow. Phosphorus, either liquid or solid, reacts with oxygen to form small amounts of two short-lived molecules, HPO and P_2O_2. Both molecules emit visible light as they are formed. The glow persists as long as new molecules are formed.

THE COMPOUNDS OF PHOSPHORUS

When phosphorus is burned with excess oxygen, the compound phosphorus oxide (P_4O_{10}) is formed. When exposed to water, it forms phosphoric acid (H_3PO_4). Phosphoric acid is used in fertilizers, detergents, food flavoring, and pharmaceuticals. Phosphoric acid is prepared commercially by heating calcium phosphate rock with sulfuric acid.

Phosphoric acid is used to create many phosphate compounds, such as triple superphosphate fertilizer ($Ca(H_2PO_4)_2 \cdot H_2O$). Trisodium phosphate (Na_3PO_4) is used as a cleaning agent and as a water softener. Calcium phosphate ($Ca_3(PO_4)_2$) is used to make china and in the production of baking powder ($NaHCO_3$).

Phosphorus is also crucial to living organisms because phosphorus compounds are used to store energy. A substance called adenosine triphosphate, or ATP ($C_{10}H_{16}N_5O_{13}P_3$), is the energy transport molecule in plants

The tusks of these walruses are covered with a hard layer called enamel. The enamel of tusks and teeth is chiefly made of a complex molecule called apatite, which contains phosphorus.

A CLOSER LOOK
UNDERSTANDING EUTROPHICATION

Phosphorus is an important plant nutrient, and in lakes and rivers it is often the limiting nutrient. A limiting nutrient is the one in shortest supply relative to other nutrients and thus prevents further plant growth. Because phosphorus is applied to farm fields and used in detergents, runoff from land often carries phosphorus with it. When the phosphorus reaches a lake, the algae and plants in the lake are able to grow quickly because of the extra phosphate. This can lead to excessive plant growth. When these plants die, the processes that break down the plant matter use up oxygen. The lack of oxygen that it causes is called eutrophication. Fish and other aquatic organisms may die as a result of eutrophication.

This drainage ditch shows the effects of phosphorus runoff. The surface of the water shows an algal bloom, where algae have flourished in the presence of unusually high amounts of phosphorus.

and animals. The human body has a limited amount of ATP. It is in constant use and constantly being recycled. Each ATP molecule is recycled 2,000 to 3,000 times every day. The human body generates, processes, and recycles about 2.2 pounds (1 kg) of ATP per hour.

THE USE OF PHOSPHORUS

Phosphates are one of the three primary plant nutrients. Fertilizers are used to supply phosphorus to plants. Some of the phosphorus in fertilizers comes from phosphate beds in sedimentary rocks. These rocks are rich in phosphates and are quarried, crushed, and added to fields. The use of the Haber process has largely replaced the quarrying of phosphate-rich rocks. In some developing countries, quarrying phosphate rocks is still more cost-effective than the Haber process. The use of phosphate rocks is still used in organic farming methods. In some areas of the world, large rafts are baited with food to attract seabirds. The bird droppings are then used as fertilizer.

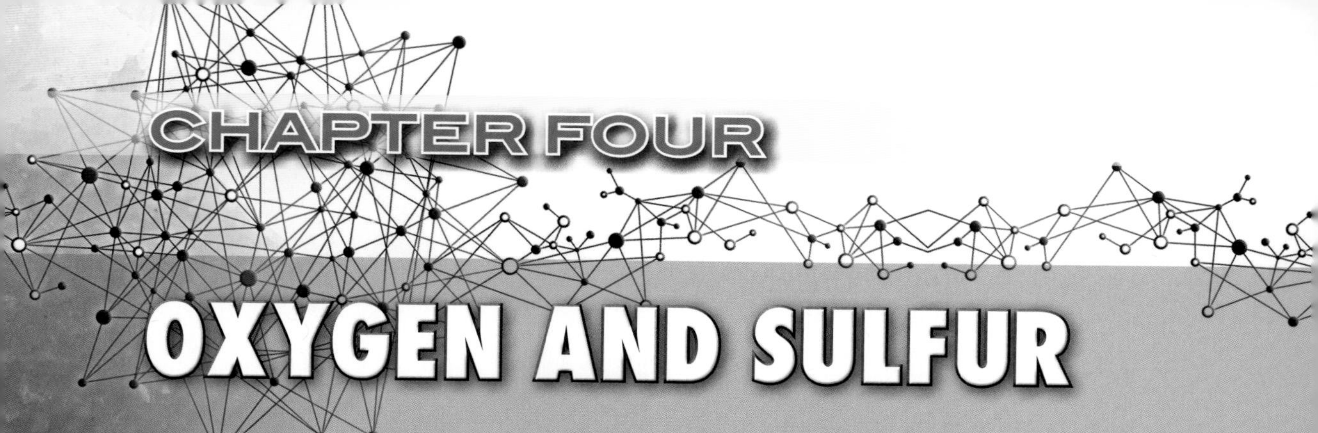

CHAPTER FOUR

OXYGEN AND SULFUR

Oxygen and sulfur are both essential elements for life. Oxygen is the second most common element in Earth's atmosphere, and it forms chemical bonds with most other elements. In nature, sulfur exists in elemental form as a yellow crystalline solid, and it also occurs in sulfide and sulfate minerals.

Oxygen and sulfur belong to Group 14 of the periodic table. Oxygen is represented by the symbol O. Oxygen has an atomic number of 8 because each atom has eight protons. Oxygen also has eight neutrons in its nucleus that, when combined with the mass of protons, gives it an atomic mass of 16. Oxygen is the second most common element on Earth. About 46 percent of the mass of Earth's crust is oxygen, and it makes up about 28 percent of the mass of the entire planet. In the universe, oxygen is the third most common element. In its diatomic state (O_2), oxygen makes up almost 21 percent of the atmosphere. The oxygen in the atmosphere comes from photosynthesis in plants and microorganisms. The oxygen is a by-product of converting

These sandstone rock formations contain many minerals. Oxygen is the most common element in Earth's crust. It forms an important part of oxide, phosphate, sulfate, silicate, and carbonate minerals.

carbon dioxide (CO_2) and water (H_2O) into glucose ($C_6H_{12}O_6$).

Sulfur is represented by the symbol S. It has an atomic number of 16 and an atomic mass of 32 (it has 16 protons and 16 neutrons in its nucleus). Elemental (uncombined) sulfur is odorless, although many people associate it with a rotten egg smell. The smell is actually from hydrogen sulfide gas (H_2S). Sulfur compounds are also responsible for giving characteristic odors to some living things, such as skunks and garlic. Sulfur is important in other ways for living organisms. Some amino acids incorporate sulfur into their structure.

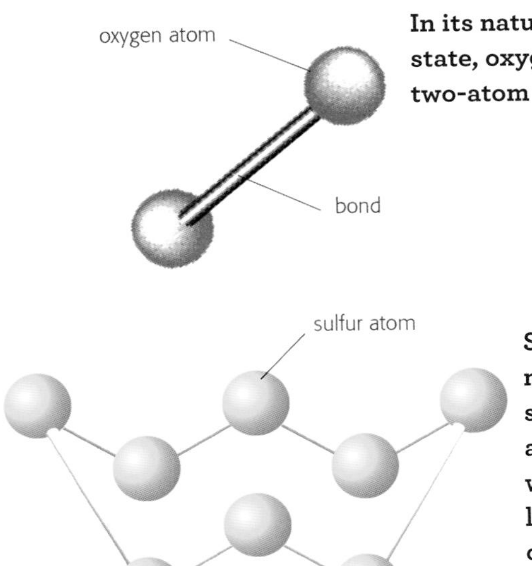

In its natural or elemental state, oxygen occurs as a two-atom molecule.

Sulfur forms a far more complicated structure of eight atoms in a ring, which has been likened to a boat or crown.

ALLOTROPES OF OXYGEN

Oxygen has two allotropes, or forms: diatomic oxygen (O_2) and ozone (O_3). Both allotropes exist in the atmosphere, but most atmospheric oxygen exists in the diatomic form. Diatomic oxygen is a more stable form than ozone. Oxygen occurs throughout the atmosphere, but ozone is usually concentrated at high altitudes. Ozone acts as a shield protecting Earth from ultraviolet radiation. It is also produced at ground level by lightning and electrical equipment. Ozone is considered a pollutant when it forms at ground level.

This aquatic plant has bubbles of oxygen attached to its leaves. Plants use sunlight, carbon dioxide, and water to produce chemicals for growth. Oxygen is a by-product of a biological process called photosynthesis, which releases oxygen into the air.

A CLOSER LOOK
THE OZONE LAYER

In the upper atmosphere, ozone is important because it filters out harmful energy rays from the Sun. A group of chemicals called chlorofluorocarbons (CFCs) destroys ozone. CFCs are used in refrigeration and cooling systems. However, if they are released into the atmosphere, they rise up to the ozone layer and react with the ozone. The ozone is converted into diatomic oxygen. Each year, a "hole" appears in the ozone layer over Antarctica. Scientists hope the hole will close now that fewer CFCs are being used.

A scientist launches a balloon over Antarctica. The balloon contains measuring devices that will record levels of ozone high in the atmosphere.

OXYGEN'S CHEMICAL PROPERTIES

Oxygen has six electrons in its outer shell and has a high electronegativity, which makes it strongly attractive for free electrons. To fill its outer shell, an oxygen atom needs to gain two electrons. Because of the small size of the oxygen atom, it readily forms double bonds. At standard temperature and pressure, an oxygen atom bonds to another oxygen atom and forms a diatomic gas. Oxygen also reacts easily with almost all other elements. When other elements react with oxygen, they become oxidized. One of the most familiar oxidation reactions is between iron and oxygen. This forms iron oxide or rust. Almost all metals react with oxygen to form metal oxides.

When oxygen forms compounds, oxygen has a negative oxidation state because it has two half-filled outer shells. When these shells are filled, the oxide ion O_2- is created. Oxygen also forms peroxides. A peroxide contains the $O_{22}-$ ion; it is assumed that each oxygen has a charge of –1.

The ability of oxygen to accept electrons by partial or complete transfer defines an oxidizing agent. The change in oxidation state from 0 to –2 is called reduction. The term *oxidation* is applied to any substance that readily accepts electrons, and oxygen accepts electrons very readily.

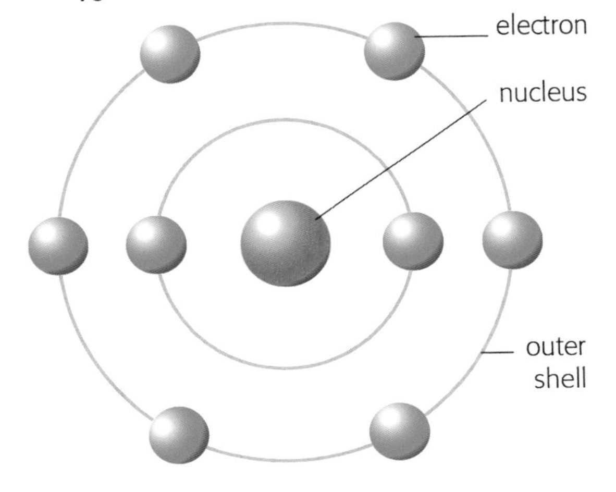

DISCOVERING OXYGEN

Most people believe the English chemist Joseph Priestley (1733–1804) discovered oxygen in 1774; he published his results in the same year. He made oxygen by heating mercuric oxide (HgO). Priestley performed further studies on oxygen and found that plants also produced oxygen.

In fact, another chemist, the German Carl Wilhelm Scheele (1742–1786), had already discovered oxygen in 1772. Scheele found that he could generate oxygen by heating up several different chemicals. However, he did not publish his results until 1777, so credit for the discovery is usually given to Priestley. Antoine Lavoisier named oxygen in 1775.

REACTIVITY

Oxygen gas does not react with itself or with nitrogen under normal conditions. In the upper atmosphere, ultraviolet radiation (high-energy rays) from the sun provides enough energy to cause oxygen, O_2, to form ozone, O_3. The ozone is then able to absorb more ultraviolet radiation and prevent these rays from reaching Earth's surface.

Oxygen is very reactive with most other elements, but it does not react with water. Oxygen will dissolve in water to

A CLOSER LOOK

CARL WILHELM SCHEELE (1742–1786)

Scheele was a Swedish chemist and pharmacist. His work as a chemist led him to discover the elements oxygen, nitrogen, barium, chlorine, manganese, molybdenum, and tungsten. He also discovered several chemical compounds, including hydrogen cyanide, hydrogen fluoride, citric acid, hydrogen sulfide, and glycerol. Scheele published only one book, in which he described oxygen and nitrogen in 1777. Scheele may have died from mercury poisoning as a result of his experiments.

a limited extent. Fish and other aquatic organisms can remove oxygen from water by diffusion (diffusion is the transport of molecules by mixing). Oxygen does not normally react with halogens, and it does not react with acids or bases under normal conditions.

ESSENTIAL OXYGEN COMPOUNDS

Oxygen forms many important compounds. One of the most familiar oxygen-containing compounds is dihydrogen oxide, or water. The chemical formula for water is H_2O. Water is a very stable molecule, and it is not easy to split it into hydrogen and oxygen. One of the best methods for splitting water is electrolysis (passing an electric current through it). When water undergoes electrolysis, it produces about twice as much hydrogen as oxygen. Oxygen also forms a stable compound with carbon. This compound is carbon dioxide, CO_2. Carbon dioxide is formed by combustion reactions and decomposition reactions. Carbon dioxide is very stable. The main decomposition of carbon dioxide takes place by plants during photosynthesis.

Fish are able to take in oxygen for respiration through their gills. The oxygen then passes into the bloodstream. Air is bubbled into fish tanks to replace the oxygen used by the fish and plants.

Rust has formed on this wrench. Rust forms when iron compounds come into contact with oxygen and water.

Oxygen reacts with many different elements and forms ions. Common ions containing oxygen include chlorates (ClO_3–), perchlorates (ClO_4–), chromates (CrO_{42}–), dichromates (Cr_2O_{72}–), permanganates (MnO_4–), and nitrates (NO_3–). Most of these ions are also strong oxidizing agents. Most metals also bond with oxygen to form oxides. Iron oxide is commonly called rust. Other oxides on the surface of metals are referred to as corrosion. These reactions happen spontaneously in air, but they may be accelerated by reduction–oxidation reactions taking place with the metals.

Oxygen also reacts with carbon compounds to form a variety of organic chemicals. These include alcohols (R–OH), aldehydes (R–COH), and carboxylic acids (R–COOH), where R denotes an organic group. Many of these organic compounds are very reactive because the oxygen easily gives up a hydrogen ion.

OXYGEN PREPARATION

In the laboratory, oxygen may be prepared by the decomposition (breaking down) of almost any oxygen-containing compound. This process is useful for obtaining small amounts of oxygen. Compounds decompose at different

CHEMISTRY IN ACTION

DESTROYING BACTERIA

Ozonation is a water treatment process that destroys bacteria and other microorganisms by bubbling ozone gas through water. Ozone acts as a strong oxidizer and kills the bacteria and other microorganisms. Ozonation is very effective for treating water, and it has an added benefit of not spoiling the taste, unlike chlorine. Also, ozone reduces the formation of trihalomethanes, compounds that form between chlorine and some organic molecules. Some scientists believe trihalomethanes may cause certain types of cancers.

Elemental sulfur forms needle-like crystals that hang in curtains. In its crystalline state sulfur has none of the odors usually associated with this element.

temperatures. Those that decompose at lower temperatures are easier to use than those that decompose at higher temperatures.

Electrolysis of water is also a method for obtaining oxygen gas. Electrolysis can be performed on a small scale in the laboratory or on a much larger scale in a factory. Large-scale electrolysis requires considerable electricity, so it is not commonly used. The most effective and usual method of obtaining oxygen is through the cryogenic (freeze) distillation of liquid air. This method produces ultrapure air. Air is cooled, and all the water vapor and carbon dioxide are removed. The air is then compressed and chilled through a number of steps until it is liquefied. The different gases are distilled from the liquid air. This process yields liquid oxygen as well as liquid nitrogen and liquid argon.

SULFUR'S HISTORY

Sulfur has been known and used by people for at least 4,000 years. It is a very distinctive yellow color and is commonly found in its elemental state around both active and extinct volcanoes. Sulfur has been used in many religious rites and medicinal treatments throughout its long history.

The Greeks and Romans used sulfur as an insecticide (a chemical that kills insects) and for making fireworks. The Romans mixed sulfur with tar, resin, bitumen, and other combustibles to make flaming bombs.

The Chinese learned to make gunpowder with sulfur during the 9th century. The use of gunpowder spread through Asia to the Middle East and finally to Europe. At first, gunpowder was used to make fireworks, and later it was adapted to make weapons.

It was not until about 1777 that the French chemist Antoine Lavoisier (1743–1794) finally convinced the scientific community that sulfur was an element and not a compound.

Until the late 1880s, the usual method to obtain sulfur was to exploit deposits that were easy to dig from the ground. Then the Frasch method was devised. This process took advantage of sulfur's low melting point. Steam was forced into sulfur deposits, which melted and were forced out of the ground.

A CLOSER LOOK

AIR POWER

Liquid oxygen, obtained by the cryogenic distillation of air, is often used as the oxidizing agent in rocket engines, which must burn large quantities of fuel very quickly. When burned with hydrogen, oxygen produces an enormous amount of energy with water as a by-product. The oxygen generator pictured is being used on board the USS *George Washington* to provide liquid oxygen for jets. Pure liquid oxygen is pale blue.

Grand Prismatic Springs in Yellowstone National Park, Wyoming. The bright colors around the edges of the water are produced by bacteria and cyanobacteria that thrive in the very hot, sulfur-rich water. Organisms that live in very hot water are called thermophiles.

SULFUR'S CHEMICAL PROPERTIES

Like oxygen, sulfur has six electrons in its outer shell. The electronegativity of sulfur is less than that of oxygen, so it is less reactive. However, sulfur is an important part of many minerals, since it readily forms sulfides and sulfates with metals. Sulfur also reacts with oxygen and can be oxidized to sulfur dioxide (SO_2).

When sulfur burns in oxygen, it produces a blue flame. Sulfur is a solid at standard temperatures and pressures. Elemental sulfur is insoluble (does not dissolve) in water, but will dissolve in carbon disulfide, CS_2. Because of its atomic structure, sulfur can either accept or lose electrons when it forms bonds. The common oxidation numbers of sulfur are −2, +2, +4, and +6. This variability allows sulfur to react with most other elements to form stable compounds.

THE COMPOUNDS OF SULFUR

When hydrogen sulfide dissolves in water, it forms an acid. This acid reacts with many different metals to form sulfides. These metal sulfides are common. One common sulfide is iron sulfide, FeS_2. Iron sulfide forms a common mineral called iron pyrite. Iron sulfide has a cubic crystal shape and a shiny, gold-colored luster. The appearance gives

When sulfur is burned in the presence of oxygen, it produces a bright blue flame

iron pyrite its common name of fools' gold. Lead sulfide, PbS, forms the mineral galena. Galena crystals were once used to control the flow of electrical current in radio sets (the crystals were used as semiconductors).

Sulfur has several oxides, which react with water to form acids. These acids react with metals to form many common sulfates and sulfites. The most common acid formed this way is sulfuric acid, H_2SO_4. Sulfuric acid is used in many industrial

CHEMISTRY IN ACTION

SMELLING SULFUR

Sulfur is usually associated with a rotten egg smell. However, that smell is not from elemental sulfur but is instead from hydrogen sulfide, H_2S. Hydrogen sulfide occurs naturally in petroleum, volcanic gases, and hot springs. Sulfur compounds called thiols have even stronger odors. Skunks secrete a mixture of several thiols to drive away enemies. Thiols are complex compounds of carbon, hydrogen, and sulfur. A skunk's powerful odor can be removed by oxidizing the thiols. Sodium bicarbonate (baking soda) is an effective oxidizing agent.

The *Rafflesia* flower stinks of rotting flesh. The odor comes from thiols and attracts the flies that pollinate it.

chemical processes. It also is used in many industrial processes, including reprocessing, fertilizer manufacture, oil refining, wastewater processing, and chemical synthesis.

Sulfur also forms many different organic compounds. In large amounts, it is toxic to living organisms, but small amounts only are required for many different compounds. Most sulfur compounds have a pungent smell. One group, the thiols or mercaptans, is used to give odorless natural gas or methane a smell. The smell is important so people know if there is a gas leak. One of the most familiar thiols is produced by skunks for their defensive spray. Sulfur compounds also give grapefruits, garlic, onions, boiled cabbage, and rotting flesh their strong, distinctive odors.

SULFUR PREPARATION

The most important sulfur compound used in industry is sulfuric acid. Sulfuric acid is used in many different industries. The production of sulfuric acid is itself a large industry. In the United States, more sulfuric acid is produced each year than any other industrial chemical. Concentrated sulfuric acid is stable at a concentration of about 98 percent. At this concentration, the acidity

Iron pyrite crystals are found in many rocks. Iron pyrite is made of iron sulfide, FeS_2. It is sometimes called fools' gold.

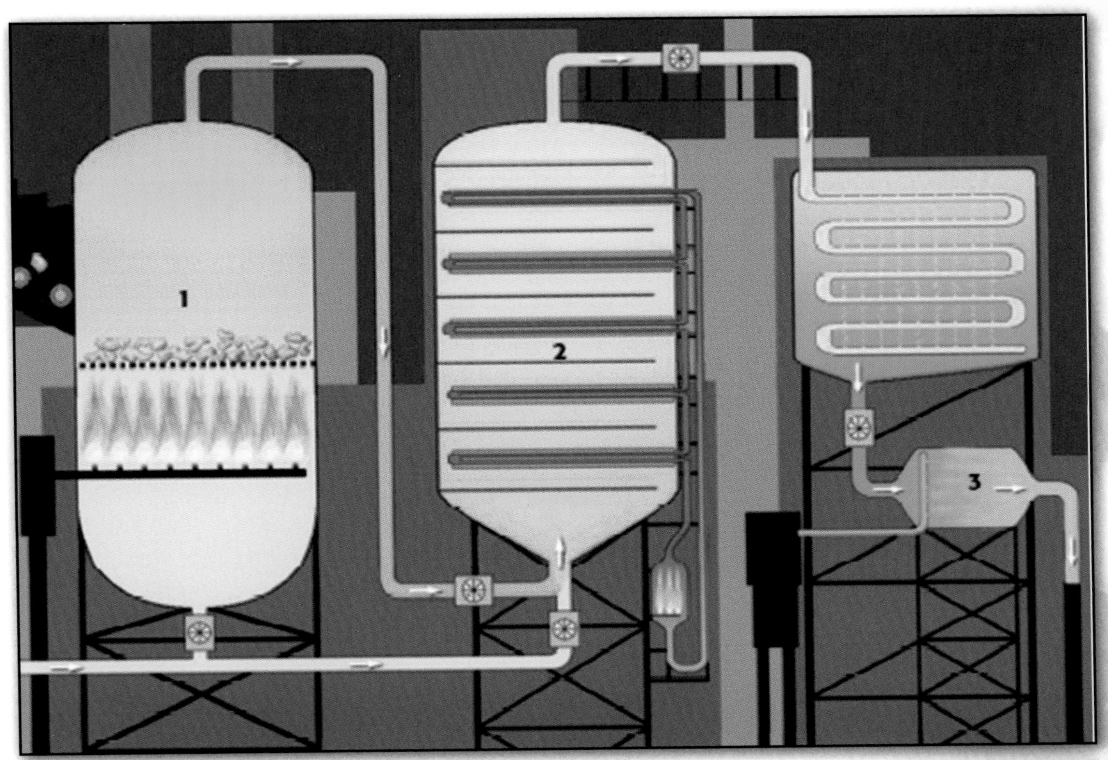

The contact process is used to make sulfuric acid. Sulfur (yellow) enters the roasting tower (1) on a conveyor belt. There, the sulfur is burned to produce sulfur dioxide gas (orange). The sulfur dioxide is piped to the oxidizing tower (2), where it is oxidized to sulfur trioxide (yellow) in the presence of vanadium oxide (red). The sulfur trioxide is pumped into another tank (3). There, water (blue) is added to make sulfuric acid.

(pH) is about 0.1. Sulfuric acid is not stable at a 100 percent concentration.

Sulfuric acid is prepared industrially from sulfur, oxygen, and water through the contact process. In the first stage of the process, sulfur is burned to produce sulfur dioxide gas:

$$S + O_2 \rightarrow SO_2$$

The sulfur dioxide is then oxidized to sulfur trioxide (SO_3), using oxygen in the presence of a vanadium(V) oxide catalyst:

$$2SO_2 + O_2 \rightarrow 2SO_3$$

Finally, the sulfur trioxide is treated with water to produce sulfuric acid at a concentration of 98 percent:

$$SO_3 + H_2O \rightarrow H_2SO_4$$

Although sulfuric acid does not burn, it presents many hazards. Even the fumes of sulfuric acid will corrode metal.

In addition, when sulfuric acid comes into contact with a metal it produces

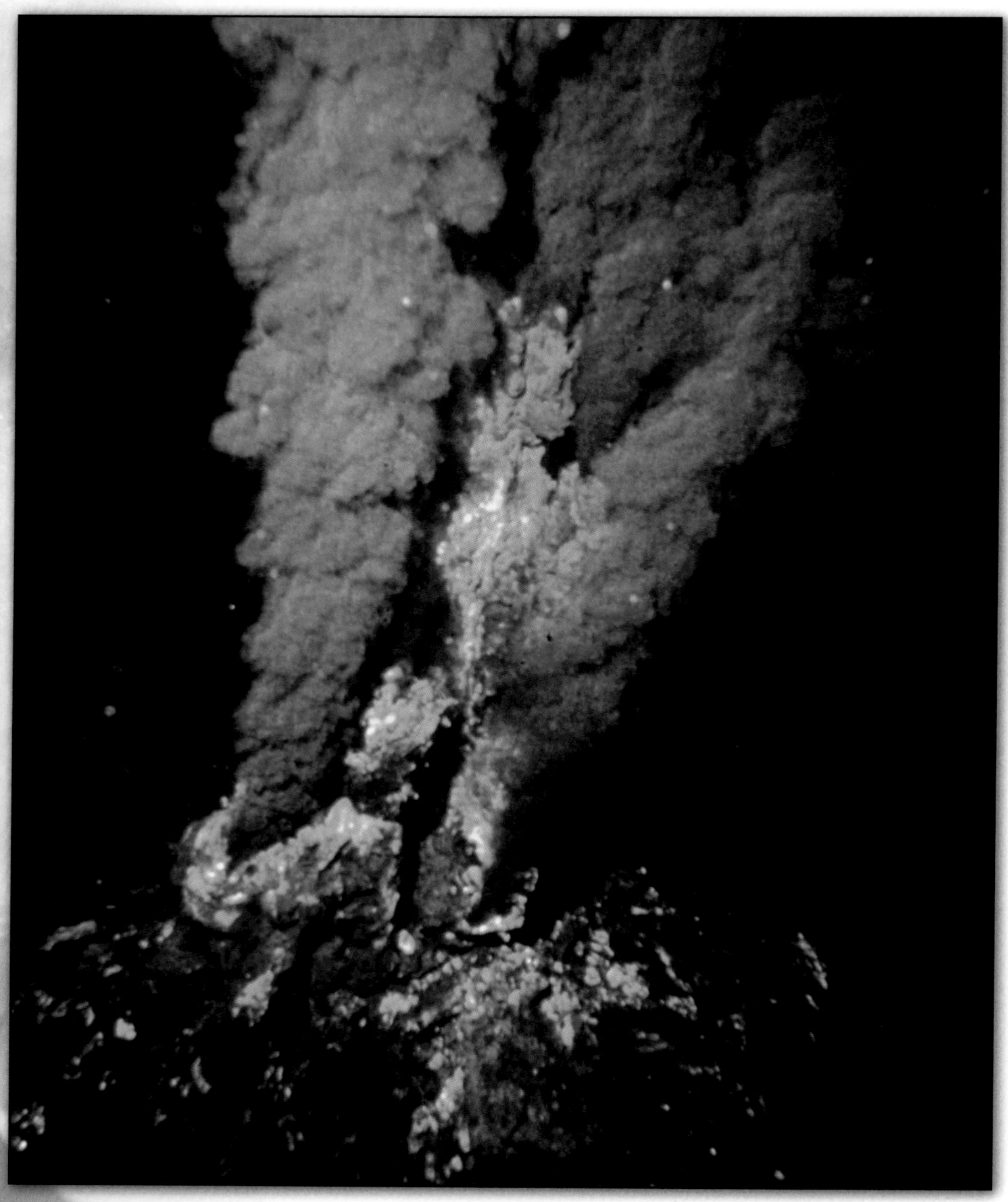

Superheated water containing dissolved sulfides pours out of a hydrothermal vent, or black smoker *(center)*. The chemicals in the water have separated out and formed a series of rock formations.

hydrogen gas, which burns. Strong concentrations of the acid also burn skin. As an aerosol, it may burn eyes.

VOLCANO VENTS

Sulfur compounds commonly occur in and around volcanic sources. Volcanoes often have vents that deposit elemental sulfur. Some volcanic vents are found deep in the ocean near rift zones. These are called hydrothermal vents, and they release superheated water laden with metal sulfides. Often, the superheated water that is released from the vent is black, so these vents are called black smokers. Over time, metal sulfide deposits may form around the vents.

Many black smokers are more than 4,900 feet (1,500 m) deep in the ocean. Their mineral deposits are not the only fascinating thing about them. They support a vast array of life such as giant tubeworms, clams, and shrimp—in fact, an entire ecosystem. Almost all other ecosystems depend on the sun's energy, which helps plants to photosynthesize (convert carbon dioxide and water to glucose). Animals then feed on the plants and use the glucose as food.

However, no sunlight ever reaches the hydrothermal vents because they are so deep underwater. Bacteria break down the hydrogen sulfide in the water and use it as the energy source to produce food. This process is called chemosynthesis. Scientists study these ecosystems to find clues to how life started on Earth. — and where life may be found on other planets.

CHAPTER FIVE

THE HALOGENS

The halogens are a highly reactive set of elements that usually occur combined with other elements. They are all colored, and vary from gases to solids.

The halogens are the elements in Group 17 of the periodic table. The halogens are fluorine (F), chlorine (Cl), bromine (Br), iodine (I), and astatine (At). The name *halogen* comes from the Greek words for "salt former." The halogens react with metals to form salts.

PHYSICAL PROPERTIES OF THE HALOGENS

In their elemental (uncombined) form, all the halogens exist as diatomic (two-atom) molecules, F_2, Cl_2, Br_2, I_2, and At_2. Fluorine is a pale yellow gas, and chlorine is a greenish yellow gas. Bromine is a

Many companies add chlorine to the water supply to kill any bacteria that might enter through a cracked pipe. The chlorine is almost completely gone by the time the water reaches the faucet, making it safe to drink.

CHEMISTRY IN ACTION

BRIGHT LIGHTS

Halogen lamps are very bright lights. They use metal halides, chemical compounds made of a metal and a halogen, to produce a bright light similar to daylight. Halogen lamps have a tungsten filament inside a quartz-glass envelope. When the lamp is turned on, the tungsten filament begins to vaporize, and the vapor reacts with halogens in the envelope. Tungsten halide is deposited on the filament. This process allows the filament and the lamp to have a longer life than a regular lamp.

Halogen lamps can be smaller than conventional bulbs because they give off a very bright light.

dark red-brown liquid that forms a red-brown gas, while iodine is a dark gray solid that forms a violet vapor when heated. Astatine is a very rare element and is radioactive. All of the halogens are poisonous in their elemental form.

The physical properties of the halogens change down Group 17. Chlorine is a gas at room temperature. Bromine is a liquid that changes easily to a gas with a slight increase in temperature. Iodine is a solid that requires heat to vaporize.

CHEMICAL PROPERTIES OF HALOGENS

The halogens react with most metals and with many nonmetals. All the halogens are highly reactive because they have a strong affinity (attraction) for electrons. All of the halogens have an oxidation number of −1, in which they try to gain one electron to fill their outer shell and make their atoms stable.

COMMONLY FOUND HALOGENS

Fluorine is the most reactive element in the periodic table. Its reactivity makes fluorine a very corrosive gas. Fluorine is fairly common in Earth's crust and forms a number of minerals. The mineral fluorite, CaF_2, is the common source. Fluorite has cubic crystals that may be colorless, white, purple, blue, green, yellow, or red. Fluorine is used to produce chlorofluorocarbons (CFCs) for refrigerants, hydrofluoric acid, steel production, and plastics such as Teflon.

Teflon is a halogenated plastic that is used on pan linings to prevent food from sticking while it cooks. Teflon can withstand high temperatures but loses its effect if the surface is damaged.

Chlorine is the most commonly used halogen in industry. The mineral halite, NaCl (common salt), is the main natural source of chlorine. Chlorine gas is used to disinfect water, chlorine compounds are used as bleach, and chlorine is used in some plastics such as PVC. Chlorine is also used in many pesticides.

Bromine and iodine are less abundant than fluorine or chlorine. Consequently, they have fewer industrial uses. Bromine is used to produce pesticides, fire retardants, and some photographic films. Iodine is an important element for human health. It is often added to table salt to prevent a hormonal condition called goiter, which affects the thyroid gland in the neck.

HALOGEN REACTIVITY

All the halogens have seven electrons in their outer shell. To fill their octet, they need only to accept one electron. Thus, the halogens are all very similar in their reactivity. They all oxidize (take electrons from) metals to form halides. The halogen oxides and the hydrides form acids in water. Fluorine is the most electronegative (has the strongest negative charge) of all elements. Generally, electronegativity and oxidizing ability become weaker from fluorine to iodine. The result of this decreasing electronegativity is increased covalent (shared) bonding in the compounds, so that aluminum fluoride (AlF_3)

Photographic films use silver halides to produce pictures. Silver halides are very sensitive to light and decompose to leave black silver grains that form the image. The process liberates the halide as free bromine, chlorine, and iodine.

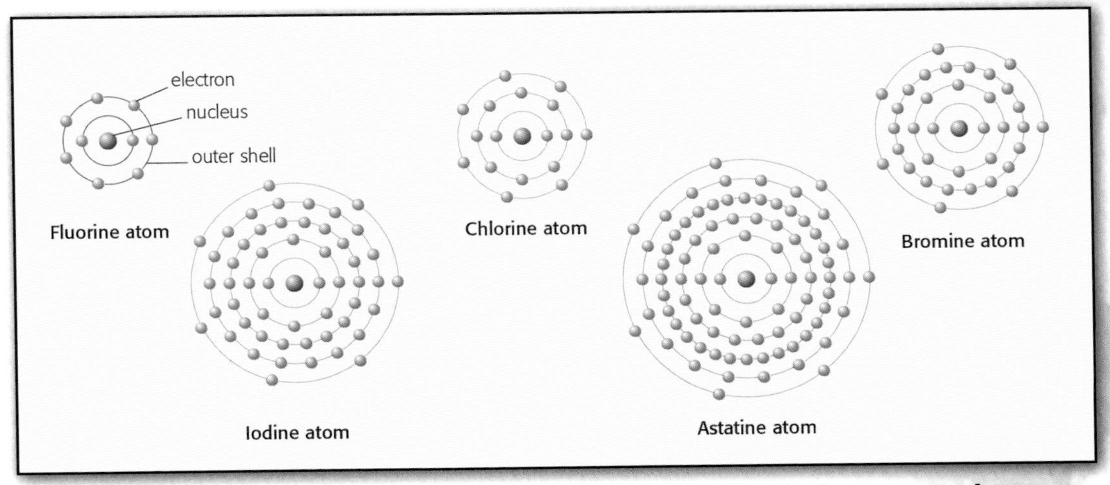

The halogens are shown ranked in ascending order of atom size. As the atoms get larger, there is less pull on the outer electrons by the nucleus. Thus they are more likely to form covalent bonds with other atoms.

is ionic (forms positively or negatively charged atoms called ions) whereas aluminum chloride ($AlCl_3$) is covalent.

Fluorine shows some peculiarities because of the small size of its atom and ion. This allows several fluorine atoms to pack around a different central atom, as in aluminum hexafluoride ($AlF6_3-$) compared with aluminum tetrachloride, $AlCl_4-$. The F–F bond is also unexpectedly weak because the small size of the fluorine atom brings the unbonded electron pairs closer together than in other halogens, and repulsion from the electrons weakens the bond.

CHANGE IN ATOMIC RADIUS

The halogens increase in atomic radius going down the group. For each halogen, the outer electrons feel a net pull of 7+ from the nucleus. The positive charge on the nucleus is cut down by the negativity of the inner electrons. Therefore, the

CHEMISTRY IN ACTION
HYDROFLUORIC ACID

Hydrofluoric acid, HF, is so reactive that it will attack glass. As a result, the acid must be stored in polyethylene or Teflon containers. Hydrofluoric acid is also very dangerous to handle. It easily penetrates the skin and destroys the underlying tissue. Hydrofluoric acid may also remove the calcium from bones. Hydrofluoric acid is widely used in the semiconductor industry to remove oxides from silicon.

number of layers of inner electrons is the only factor that affects the size of the atom.

Electronegativity is a measure of the tendency of an atom to attract a bonding pair of electrons. The most electronegative element is fluorine. The bonding pair of electrons between a hydrogen atom and a halogen feels the same net pull of 7+ from both fluorine and chlorine. The stronger pull from the closer fluorine nucleus is why fluorine is more electronegative than chlorine. As the halogen atoms get larger, any bonding pair gets farther and farther from the halogen nucleus, and so it is less strongly attracted toward it. Thus, the electronegativity decreases.

Calcium fluoride (CaF_2) occurs as the mineral fluorite. It is one of the primary sources of fluorine.

Antoine Balard was a French chemist who discovered bromine. He also investigated the chemistry of how chlorine bleaches colored compounds.

The electron affinity is a measure of the attraction between the incoming electron and the nucleus. The higher the attraction, the higher the electron affinity. The trend down Group 17 is for the electron affinities to decrease. As the atom gets larger, the incoming electron is farther from the nucleus and so feels less attraction. The electron affinity therefore falls as you go down the group. In the example of fluorine, because the atom is very small, the existing electrons are all close together. Therefore, the repulsion they produce is particularly great. This lessens the attraction from the nucleus enough to lower the electron affinity below that of chlorine.

DISCOVERING HALOGENS

Fluorite, CaF_2, was first described in 1530. It was used to help join metals. Many famous early chemists experimented with hydrofluoric acid, which is obtained by treating fluorite with concentrated sulfuric acid. It was known that hydrofluoric

CHEMISTRY IN ACTION

A COMMON HOUSEHOLD CLEANER

Bleach is commonly used in the laundry. A little bleach makes whites even whiter, while too much bleach will turn colored clothes pale. Household bleach is a dilute solution of sodium hypochlorite. Bleach is also widely used throughout many different industries. Bleach eliminates surface germs, bacteria, and viruses. Bleach is used to disinfect food-processing equipment and hospital equipment. It is also used to disinfect swimming pools. Industries add it to cooling water to prevent the growth of bacteria and algae in pipes. Sodium hypochlorite is also used in recovering precious metals during mining, and to bleach wood pulp during the manufacture of paper.

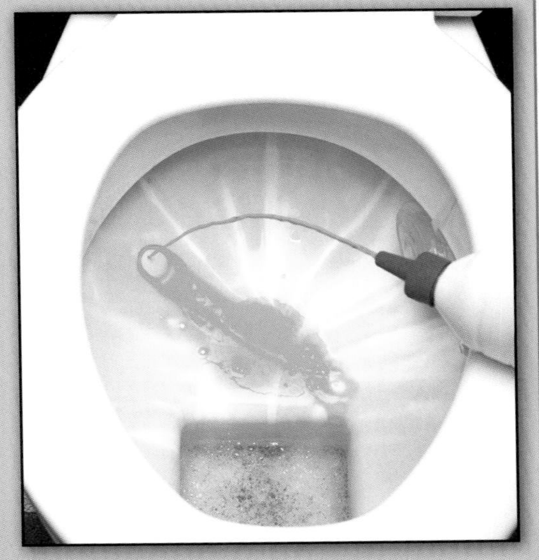

Toilet cleaners contain bleach to kill germs. It is important not to mix bleach with other cleaning materials because this may produce a dangerous reaction.

acid contained a new element, but it could not be isolated because of its high reactivity. In 1886, the French chemist Henri Moissan (1852–1907) isolated fluorine. He was awarded the Nobel Prize for his discovery in 1906.

Swedish chemist Carl Wilhem Scheele discovered chlorine in 1774, but he mistook it for oxygen. The British chemist Humphry Davy (1778–1829) finally managed to isolate chlorine in 1810.

Antoine Balard (1802–1876) discovered bromine in 1826. He extracted it from salt-marsh soil in France. Joseph-Louis Gay-Lussac (1778–1850) suggested the name (from the Greek *bromos* meaning "stench") because of the characteristic smell of the vapors.

Sodium fluoride is added to some toothpastes to help strengthen teeth.

Barnard Courtois (1777–1838) discovered iodine in 1811. He was extracting saltpeter (potassium nitrate) from seaweed ash using concentrated sulfuric acid. He accidentally

A CLOSER LOOK

THE CHEMISTRY OF CHLORAMINES

Chloramines are compounds that contain nitrogen, hydrogen, and chlorine. They sometimes occur when chlorine-containing household cleaners are accidentally mixed with ammonia in other cleaners. The reaction produces chloramine gas. Exposure to chloramine causes irritation to the eyes, nose, throat, and airways. Symptoms include teary eyes, runny nose, sore throat, coughing, and chest congestion. These symptoms may develop after only a few whiffs of chloramine and may last up to 24 hours. Chloramines are responsible for the characteristic smell at indoor swimming pools.

KEY DEFINITIONS

- **Ion:** An atom or molecule that has gained an electric charge through loss or gain of electrons in its outer shell.
- **Oxidizer:** A substance that removes electrons from another substance to make its own outer shell stable.

added too much acid and noticed a purple vapor. He found that the purple vapor would crystallize on a cold surface. Samples made their way to Gay-Lussac and Davy. They each identified the new substance as an element only days apart. They publicly argued over who identified it as an element first, but both gave credit to Courtois for discovering it.

HALOGEN OXIDATION

An oxidizer gains electrons. The more readily it gains electrons, the stronger the oxidizer. Because the halogens readily accept an electron, they are all strong oxidizers. In oxidizing ability, the halogens follow the expected order: F_2 >(stronger than) Cl_2 > Br_2 > I_2. Fluorine reacts explosively with so many organic compounds that working with it requires special facilities.

The most commonly used halogens are chlorine and bromine. Chlorine gas can be generated in dilute solution from bleach, and bromine gas is a volatile reddish brown corrosive liquid. One compound that is used for chlorination is sodium hypochlorite, NaOCl, common household bleach. In water, part of the NaOCl is converted to HOCl, which reacts as if it were $HO-Cl^+$.

ESSENTIAL HALOGEN COMPOUNDS

Two very important halogen compounds have already been discussed: hydrofluoric acid and sodium hypochlorite. While these both have many industrial applications, they are not the only important halogen compounds.

Fluorine is very reactive, so it is no surprise that it forms many compounds. As an ion, it forms fluorides with most metals and many nonmetals. Fluorides are used in many industrial applications such as producing uranium and plastics as well as in toothpaste to strengthen teeth. With organic substances, it forms organofluoride compounds. Organofluorides are used in air conditioning and refrigeration.

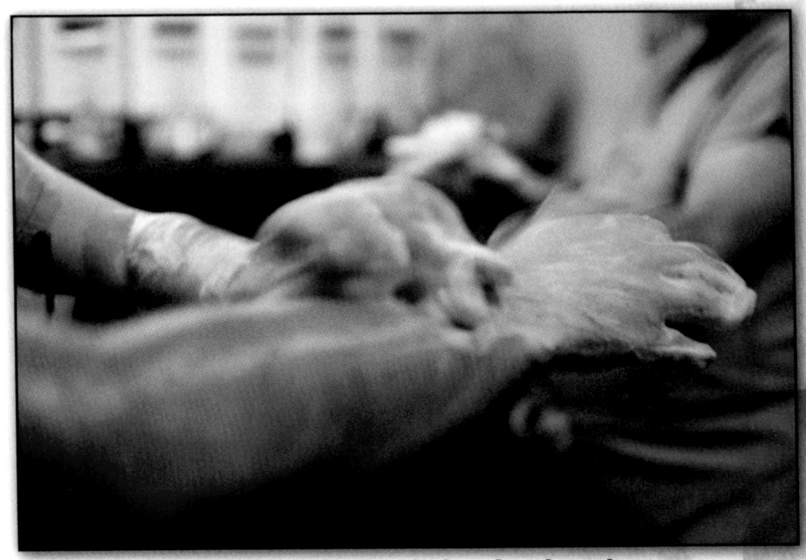

Surgeons often wash their hands with an iodine-containing soap before they begin an operation. Iodine has antiseptic properties that kill bacteria on the skin.

Like fluorine, chlorine is very reactive. Many different chlorine compounds exist. Chlorine compounds form chlorides (Cl–), chlorate (ClO_3–), chlorite (ClO_2–), hypochlorite (ClO–), and perchlorate (ClO_4–). Hydrochloric acid, HCl, is widely used in industry. Chlorine compounds are also used as oxidizers. One common use for oxidizers is as bleach. Chlorine also forms compounds with organic molecules. These organochlorides include most pesticides as well as some chemical warfare agents.

Bromine forms salts called bromates, BrO_3–. The bromates are strong oxidizers and are commonly used in fireworks. Bromates may also be formed in drinking water that contains bromide dissolved in the water where ozone has been used as a disinfectant. Bromates are carcinogenic, or cancer-causing agents. Bromide reacts with organic compounds to form organobromides.

Iodine forms salts that are iodides (I–) and iodates (IO_3–). Iodine is important to the human body and the body obtains it from iodides in food. Iodine compounds are used in photographic film and as an antiseptic for cleaning wounds or before surgery. Iodine also reacts with organic compounds to form organoiodides. These are used in medical research.

ORGANIC COMPOUNDS CONTAINING HALOGEN

Organic compounds containing halogens are called halocarbons. Halocarbons have one or more carbon atoms linked by covalent bonds with one or more halogen atoms. Some halocarbons are produced naturally when halide salts react with organic compounds. The halocarbons are, however, produced in very small amounts. The synthesis of halocarbon compounds began in the early 1800s. Today, halocarbons are used in many different products and in many different industrial processes.

Halocarbons are used as solvents, adhesives, pesticides, refrigerants, fire-resistant oils, sealants, electrically insulating coatings, plasticizers, and plastics. Halocarbons are widely used because they are very stable and very effective at what they do. Halocarbons are usually not affected by acids or bases, they many not be flammable, and they resist attacks from bacteria and molds, and many are resistant to decay from sunlight. However, these properties that make them useful also create problems.

Halocarbons are persistent in the environment, too. Pollution from halocarbons is a real problem. Because halocarbons are stable, they take a long time to break down and tend to accumulate in the environment. To avoid this buildup becoming a problem, the amount of halocarbons in industry has been reduced and the handling and disposal of halocarbons has become better regulated.

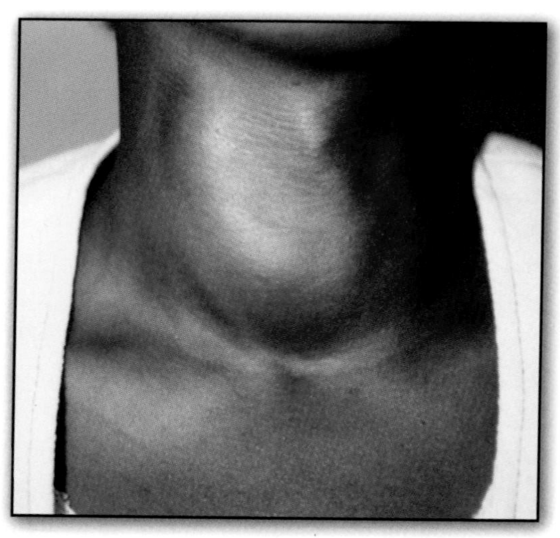

The swelling on this person's neck is called a goiter. One cause of this condition is a lack of iodine in the diet. It can be remedied by adding potassium iodide to table salt.

HALOGENS AND THE HUMAN BODY

Fluoride-containing compounds are often added to toothpaste, mouthwashes, and drinking water to help strengthen teeth

The bald eagle was once endangered by the use of halocarbon pesticides that became concentrated in the food chain.

and prevent cavities. The fluoride binds to the hydroxyapatite crystals in enamel and hardens it. This prevents tooth decay. Some water supplies have naturally occurring fluorides, and some municipalities add fluoride to the drinking water.

Chlorides are also important in the human body. In fact, chlorides make up about 0.15 percent of the total human body weight. Chloride helps the body maintain the concentrations of sodium and potassium ions in body fluids. Chloride is also important in producing hydrochloric acid in the stomach to aid in food digestion. The human body is seldom deficient in chlorides because they are found in so many different foods.

Iodine, in trace amounts, is very important to the body. Iodine makes up around 0.00004 percent of the total human body weight. The thyroid gland uses iodine to produce the hormones thyroxin and triiodotyrosine. These

CHEMISTRY IN ACTION

TOXIC GAS

Chlorine gas is very toxic and should be avoided. Many household cleaners contain hypochlorite bleach, and care should be taken when they are used. Bleach and cleaners containing bleach should never be mixed with acids. Doing so will release chlorine gas. Chlorine gas irritates the mucous membranes and may cause other medical problems such as pulmonary edema (buildup of fluid in the lungs). Read the warning labels and follow the directions when using any household cleaner.

Chlorine was the first gas to be used as a chemical weapon. Many soldiers in World War I (1914–1918) never properly recovered from its effect on their lungs during gas attacks.

hormones affect growth, development, and the metabolic rate of the body. Iodine is readily absorbed from foods, especially seafood. To ensure people get enough iodine, table salt is iodinated. This trace amount of iodine is enough to ensure people get sufficient iodine.

PREPARATION IN THE LAB

Because of the usefulness of many of the halogens, there is little need to prepare them in the laboratory in small quantities. The halogens are also difficult to produce in small quantities because they are so reactive. However, there are a few interesting methods of preparing elemental halogens.

One hundred years after fluorine's discovery, Karl Christe (1937–) discovered a new method of preparing elemental fluorine in 1986. Using this method, anhydrous (water-free) hydrofluoric acid, potassium manganese fluoride (K_2MnF_6), and antimony fluoride, SbF_3, are allowed to react in solution at 302 degrees Fahrenheit (302°F; 150°C). This method is not suitable for industrial applications but it does not require electrolysis like Moisson's process.

Adding concentrated hydrochloric acid to a sodium chlorate solution

An illustration of Henri Moisson preparing fluorine in his laboratory. His discovery led to him being awarded the Nobel Prize for chemistry in 1906.

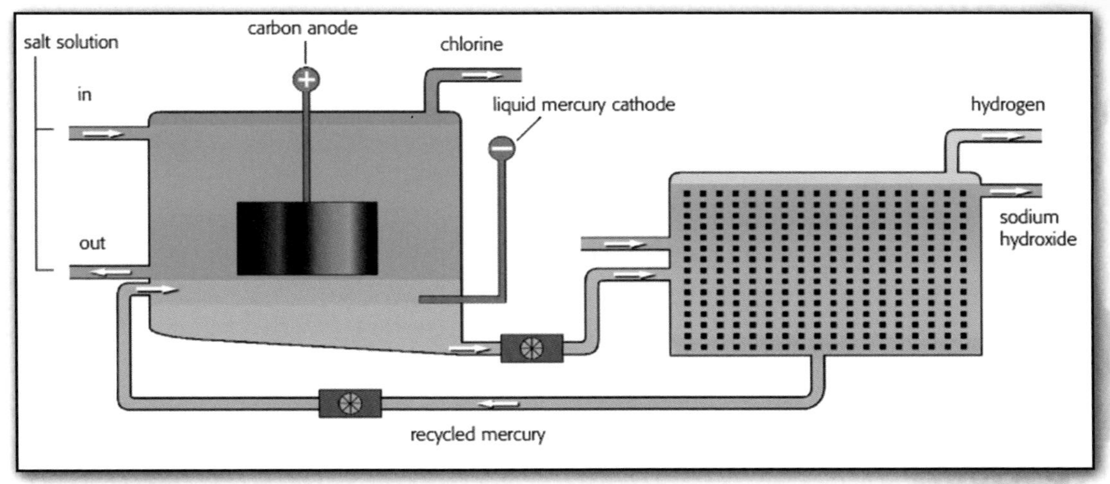

A diagram showing the chlor-alkali process for obtaining chlorine. The process works by passing an electric current through saltwater. Hydrogen gas and sodium hydroxide are also produced using this method.

produces chlorine gas. Chlorine gas is also made using other more complex reactions.

PREPARATION IN INDUSTRY

Fluorine is still produced industrially using Moisson's process. This process involves the electrolysis of anhydrous hydrofluoric acid with the addition of potassium hydrofluoride, KHF_2.

Chlorine is one of the widely used halogens. Several methods have been used to produce chlorine commercially. The most common method is membrane cell electrolysis, or chlor-alkali process. This method is used because it produces three useful industrial products: chlorine gas, hydrogen gas, and sodium hydroxide. The overall reaction for this method is:

$$2NaCl + 2H_2O \rightarrow Cl_2 + H_2 + 2NaOH$$

The chlor-alkali process takes place in a reaction cell. Chlorine forms at the anode (positively charged electrode), and sodium hydroxide and hydrogen form at the cathode (negatively charged electrode). This process is efficient and is used to produce large quantities of each product.

Herbert Dow (1866–1930), who founded Dow Chemical, discovered the electrolytic method for recovering elemental bromine from brine deposits. Brine (saltwater) deposits often occur with petroleum. The brine is sometimes very high in bromine compounds. The electrolysis of the brine produces elemental bromine, which can be used to produce commercial bromine compounds.

CHAPTER SIX

NOBLE GASES

Only one group in the periodic table consists entirely of gases. This group is called the noble gases because they appear not to react with other substances.

The noble gases are the chemical elements in Group 18 of the periodic table. This series contains helium (He), neon (Ne), argon (Ar), krypton (Kr), xenon (Xe), and radon (Rn). The elements in this

The sun is slowly turning from a fiery ball of hydrogen atoms into a mass of helium atoms. The extreme temperatures and pressures in its core push the hydrogen atoms together, where they fuse into helium. This photograph is a false-color image.

group are the least reactive of all the elements. The reason for their unreactivity is that their atoms have a full outer electron shell, which makes them very stable.

THE PHYSICAL PROPERTIES OF NOBLE GASES

All the noble gases are monatomic (exist as single atoms). They boil at low temperatures because there are only weak forces acting between the atoms. For this reason, all members of this group are gases at standard temperature and pressure. Helium has the lowest boiling point of any substance at −452 degrees Fahrenheit (−452°F; −268.9°C).

THE CHEMICAL PROPERTIES OF NOBLE GASES

This group was originally called the inert gases because it was thought they formed no compounds. Compounds of these gases were first synthesized in 1962 and are now well documented. Helium, neon, and argon form no known compounds. Krypton forms KrF_2, a colorless solid, when it reacts with fluorine. Xenon forms a wide range of compounds with oxygen and fluorine. In fact, at least 80 compounds are known.

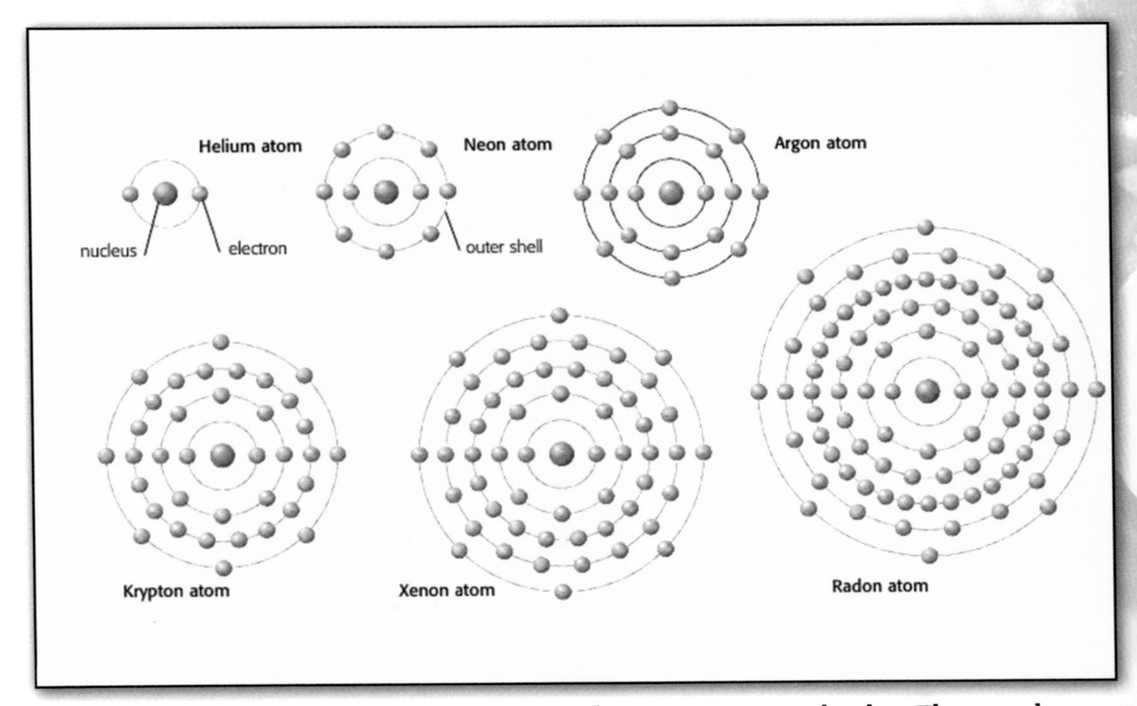

There are six noble gases: helium, neon, argon, krypton, xenon, and radon. They are the most unreactive of the elements because they have a full outer shell (containing eight electrons). Consequently, the noble gases do not need to gain or lose electrons to achieve stability.

AN EXPERIMENT LEADS TO DISCOVERY

Argon was the first noble gas discovered. Two British scientists, Lord Rayleigh (1842–1919) and William Ramsay (1852–1916) discovered it in 1894 in an experiment in which they removed the oxygen and nitrogen from air. Since argon makes up about 1 percent of air, it formed the bulk of the remaining gas. The name *argon* is Greek for "inactive."

William Ramsay discovered helium in 1895. He was looking for argon in minerals but discovered helium instead. In 1909, two other British scientists, Ernest Rutherford (1871–1937) and Thomas Royds (1884–1955), identified that an alpha particle from radioactive decay is a helium nucleus.

Ramsay and Morris Travers (1872–1961) went on to discover krypton and neon in 1898. They were investigating the components of liquefied air when they found these two gases.

CHEMISTRY IN ACTION

NEON GAS

Neon lights are tubes containing neon gas. An electric current is passed through the tube, which energizes the electrons in the neon atoms and causes them to glow red. Adding small amounts of argon, mercury, or phosphorus produces other colors. The first neon light was made in 1902, just four years after the discovery of neon.

Neon lights are frequently used to illuminate buildings at night.

Radon was discovered in 1900 by the German physicist Friedrich Ernst Dorn (1848–1916). He was studying radium and discovered radon when it formed as part of radium's radioactive decay chain. In 1908, William Ramsay and Robert Whytlaw-Gray (1877– 1958) isolated radon and determined its density. Radon is the heaviest known gas.

Lord Rayleigh in his laboratory. Rayleigh found argon while investigating liquefied air.

COMMON USES

Even though the noble gases are not reactive, some of them have commercial uses. Helium, argon, and neon are used in various applications while krypton, xenon, and radon are not. Helium is the most commercially useful of all the noble gases. Because helium has the lowest boiling point of any substance, liquid helium is used to keep things very cold. As a liquid, helium is considered a superfluid because it has no viscosity. Superfluids are used in some research instruments such as precision gyroscopes used in gravity research.

Helium is used in party balloons and lighter-than-air vessels such as airships. Helium has almost as much lifting power as hydrogen but, unlike hydrogen,

KEY DEFINITIONS

- **Radioactive element:** An element with an unstable nucleus that breaks down to form different elements.
- **Decay chain:** The sequence in which a radioactive element breaks down.

helium is not flammable. Helium is also used in industrial applications, including as a coolant in nuclear reactors and for pressurizing liquid-fuel rockets.

Argon is the gas used in incandescent lightbulbs. Argon is useful because it does not react with the filament of the bulb, even at high temperatures. It is also used as a gas shield in some forms of welding to prevent oxides from forming. Argon does not conduct heat well, so it is used between panes of glass in thermal windows and to fill dry suits worn by divers in very cold waters. Argon is also used in some museum conservation projects to prevent oxygen or water vapor in air from damaging important books and documents.

Neon is most commonly used in neon lights. However, it does have other uses. Neon is used in television tubes. It is also used in vacuum tubes, high-voltage indicators, and lightning arresters. Liquid neon is also used in some applications that do not require temperatures as cold as liquid helium.

A CLOSER LOOK

SCUBA DIVING

Scuba divers are usually restricted to shallower depths because of the effects of nitrogen and oxygen as pressure increases. Mixed-gas diving uses mixtures of helium and oxygen. By replacing the nitrogen with helium, several problems are avoided. First, at depth nitrogen can dull the senses and make the diver sleepy. Second, nitrogen dissolves in the blood. If the diver swims back to the surface too quickly, the dissolved nitrogen comes back out of solution and forms bubbles in the blood. These bubbles cause a condition called the bends, which can be fatal.

This diver is breathing a mixture of helium and oxygen. Helium is added because pure oxygen can cause dizziness, nausea, and convulsions at depth.

CHEMISTRY IN ACTION
LASERS

Helium-neon, or HeNe, lasers produce a red beam. A glass tube filled with HeNe is the body of the laser, which has two electrodes. A current is passed through the tube, and the gases become energized. A mirror at one end of the tube directs the laser beam out the other end of the tube. These lasers are used for many optical applications, such as bar-code scanners and eye surgery. Green argon lasers are also used for eye surgery. Krypton lasers are used in scientific research and light shows. Xenon lasers are used mainly in scientific research.

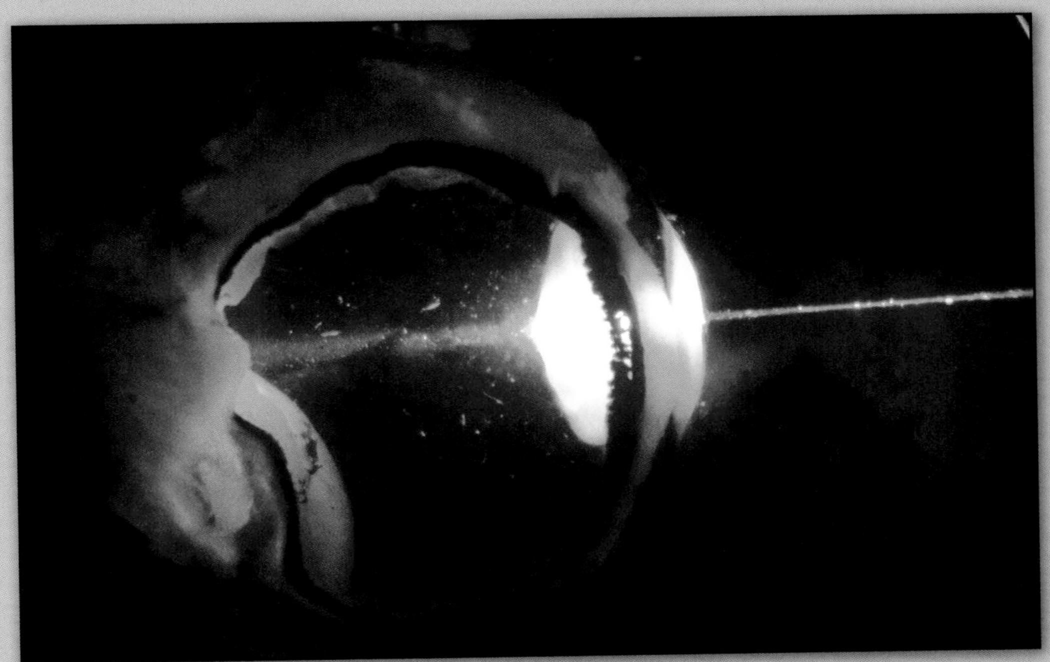

This eye is being treated using a red helium–neon laser. The beam can pass through the front of the eye to the back without the need for surgery.

INDUSTRIAL PREPARATION OF NOBLE GASES

Cryogenic (freeze) distillation is the main method of producing ultrapure noble gases. This process requires considerable energy, but it results in liquid air. Air is cooled and all the water vapor and carbon dioxide are removed. The air is then compressed and chilled through a number of steps until it liquefies. This process yields large amounts of liquid nitrogen

KEY POINTS

Concentration of noble gases in air:

Argon (Ar)	0.9340 percent
Helium (He)	0.00524 percent
Neon (Ne)	0.001818 percent
Krypton (Kr)	0.00114 percent
Xenon (Xe)	0.0009 percent

as well as liquid oxygen. The different gases are then separated out of the liquid air by raising the temperature. Each element has its own boiling point, so each gas can be collected as it changes from a liquid to a gas. Argon is the largest fraction of noble gas in air.

Helium also occurs mixed with natural gas. It can be obtained in commercial quantities through the distillation of liquefied natural gas.

ONE OF THE HEAVIEST GASES

Radon is one of the heaviest gases, with an atomic mass of 222. Radon has 20 isotopes, but none are stable. (An isotope is a version of an element's atom that has a different number of neutrons

Some types of rocks contain radioactive elements such as uranium or radium. As radioactive elements decay, they break down into other elements. Some elements form radon as they decay. Radon can escape through cracks in the rock and collect in the spaces under houses.

in the nucleus.) All radon's isotopes are radioactive and have a short half-life. Radon-222 is the most stable isotope, and its half-life is 3.8 days. Radon-222 is a decay product of radium-226 and it emits alpha particles (helium nuclei) as it decays. Radon-220 is a natural decay product of thorium and is called thoron. It has a half-life of 55.6 seconds and also emits alpha particles. Radon-219 is derived from actinium and is called actinon. It, too, emits alpha particles and has a half-life of 3.96 seconds.

Radon is usually found in soil, groundwater, and caves because these areas trap the radon. The amount of radon present depends on the amount of radioactive minerals in the ground. When in contact with the atmosphere, radon spreads out rapidly. In some areas, buildings can actually trap the radon from the soil, and high concentrations build up in basements.

Even though radon has a very short half-life, it is considered a health hazard because it causes lung cancer. While radon is cleared from the body within days, some of its decay products have much longer half-lives. It is these that cause the damage because they stay in the lungs longer. Some studies indicate that radon is the second most common cause of lung cancer after tobacco smoke.

CHAPTER SEVEN

BIOGRAPHY: HENRY CAVENDISH

This illustration depicts Henry Cavendish, who is credited with discovering the nonmetal hydrogen, the most abundant of the elements.

Henry Cavendish was born in 1731 in Nice, France, to a wealthy British family. Both his parents were children of dukes. Not much is known about Henry's youth, but his mother died when he was just two years old. Henry and his younger brother, Frederick, attended the Hackney Academy near London, a school for upper-class children. When he was eighteen, Henry's education continued at Peterhouse, a college at the University of Cambridge. He studied there for three years and left without receiving a degree, which was not unusual at that time.

After leaving Peterhouse, Cavendish returned to his family home in London. Henry's father, Lord Charles Cavendish, was a gifted scientist who encouraged his son's interests in science and invited him to assist in numerous experiments. Charles also introduced Henry to the scientific organization the Royal Society of London, taking

THE ROYAL SOCIETY

The Royal Society (its full name the Royal Society of London for Improving Natural Knowledge) is the oldest national scientific society in the world. It formed in 1660, according to its Web site, "for the promoting of Physico-Mathematicall Experimentall Learning." The organization was meant to gather knowledge, encourage research, and share findings with the public. This last aim resulted in the scientific journal *Philosophical Transactions*, which is still in publication. Existing members, called fellows, elect new members to the Royal Society.

Besides Henry Cavendish, other famous scientists have been associated with the society over the years. Isaac Newton, the mathematician and physicist, served as its president from 1703 to 1727. Benjamin Franklin presented his lightning experiment findings in a paper to the Royal Society in 1752. In 1921, Albert Einstein was elected to membership. There are currently 1,450 fellows and foreign members of the Royal Society, including physicist Stephen Hawking.

The Royal Society of London for Improving Natural Knowledge is located in London, England. Today, it is the leading independent scientific body of the United Kingdom and the Commonwealth.

him to meetings beginning in the late 1750s. Henry became a member himself as well as of other organizations, such as the Royal Institution of Great Britain and the British Museum. Throughout his lifetime, Henry Cavendish published only eighteen papers, a fact that belies his vast research and achievements. He had wide interests, and explored optics, electricity, mathematics, astronomy, and aeronautics, besides most famously chemistry and physics.

When zinc is placed in hydrochloric acid, hydrogen gas is produced.

In 1766, Cavendish's first publication—"Three Papers Containing Experiments on Factitious Airs"—focused on the properties of gases. "Factitious air" was the name he used for "any kind of air which is contained in other bodies in an unelastic state, and is produced from thence by art," or, in other words, gases given off by substances through chemical reactions. In one of these experiments, Cavendish observed that when the metals zinc, iron, and tin were submerged in certain acids, a gas was produced. We call this gas hydrogen, but he called it "inflammable air," noting its readiness to burn. Cavendish collected the hydrogen produced by the reaction in glass bottles, measured it against common air, and found hydrogen to be much lighter. Cavendish's study of hydrogen and his recognition that it was a unique gas have given him the title "discoverer of hydrogen." The remainder of Cavendish's publication focused on his study of "fixed air," later called carbon dioxide, as well as gases produced by decay and fermentation. Though he wasn't the first to do so, Cavendish's isolation of different gases further discredited the idea that air was a single element. Instead, air is composed of more than one gas.

Cavendish's next paper, published in 1783, focused on eudiometry, which is the measurement of the goodness of the air. He found that there was not much difference between samples of air taken at different times and in different places. Samples were even collected at various heights via air balloon. This study offered proof that Earth's atmosphere has a fairly constant composition.

James Watt made vital improvements to the steam engine, contributing to the Industrial Revolution.

Cavendish's next breakthrough produced water from gases. Previously, British scientist Joseph Priestly had performed an experiment in which he introduced an electric charge to hydrogen and common air. A small amount of dew resulted. Cavendish decided to explore why this would happen. In his paper "Experiments on Air" (1784), Cavendish explained that when he burned "dephlogisticated air" (oxygen) with "inflammable air" (hydrogen) in a glass container, dew appeared. His study essentially proved that water is not a single element but composed of the elements oxygen and hydrogen in the approximate proportion of two parts hydrogen to one part oxygen. Though Cavendish performed these experiments first, another scientist, James Watt, published his similar findings before Cavendish.

ANTOINE-LAURENT LAVOISIER

Born 1743, Antoine-Laurent Lavoisier became a founder of modern chemistry. In the beginnings of chemical studies, a once accepted theory was that a substance called phlogiston was contained in all matter and that the process of combustion, or burning, was a release of phlogiston. Upon the discovery that different gases make up common air, Lavoisier began experimenting with the gas later called oxygen and found that it, not phlogiston, was the vital gas of combustion and reacted with matter to produce heat and light. Lavoisier also performed experiments with hydrogen and oxygen to create water, similar to Henry Cavendish's, though Cavendish's studies were found to predate his.

In 1787, because the study of chemistry had advanced so much, Lavoisier published an improved system for naming elements and compounds. He named hydrogen based on Greek words meaning "water-former," and oxygen from Greek words meaning "acid-former." Lavoisier was executed by guillotine in 1794 during the French Revolution.

By-products of these experiments led to more interesting discoveries. One was man-made nitrates. Cavendish found that an electric charge combined water, oxygen, and nitrogen (which he called "phlogisticated air"), resulting in nitric acid. This study became the impetus for the making of nitrates used in fertilizers and explosives. Cavendish had also found that after combining oxygen and nitrogen a very small volume of gas remained. This meant another gas was unaccounted for in common air. This "leftover" gas was later studied and named argon.

Scientists today are still amazed at Cavendish's mathematical accuracy. Even though he did not have the advantage of modern scientific tools, his measurements come astoundingly close to modern numbers. For example, Cavendish's analysis of common air led him to believe that oxygen makes up about 20.8 percent of the atmosphere. Today, that number is thought to be about 20.95 percent.

Cavendish's most famous research in the realm of physics is another study of this precision. Cavendish was given an instrument called a torsion balance, which was developed by British scientist and Royal Society member John Michell. Reverend Michell had died before using the tool for its purpose: measuring the density of Earth. The balance consisted of two identical pairs of round lead weights, one large and one small. Each was suspended using a system of rods and wires.

Cavendish knew from Isaac Newton's law that all objects attract each other through the force of gravity. He hypothesized that if he measured the density of the spheres and then the force of gravitational attraction they exerted on each other (called the gravitational constant), he could use these measurements in a proportion to figure out the density of Earth.

Measuring the force of gravity between the spheres' masses involved very exact, very slight measurements. To avoid any outside factors affecting the spheres' movements, Cavendish placed the torsion balance (later called the Cavendish apparatus) in a shut room. He used a telescope and a peephole to observe it. After a year of measurements and calculations, Cavendish reached his determination that Earth's density is about 5.48 times that of water. The current accepted value is 5.52. Using Cavendish's math, Earth's mass was calculated to be about 13,000,000,000,000,000,000,000 pounds, or thirteen sextillion pounds (six sextillion metric tons). The accepted value today differs by only about 1 percent.

This is a 1:48 scale model of Henry Cavendish's torsion balance, an instrument for measuring weak forces of attraction between masses.

Much more of Cavendish's work was discovered after his death in 1810. In one report on electricity, Cavendish used himself as a galvanometer. He measured the resistance of electric current through different liquids by noting the pain he felt as well as how far up his arm it traveled. One of the results of his many experiments was that salt water is a much better conductor than freshwater. Incredibly, even these outcomes have proven accurate when compared with galvanometers of today. Other unearthed research showed that Cavendish's experiments with temperature and pressure could have led to the steam engine developing years earlier than it had.

One reason for Cavendish's hesitation to share his findings may have been his extreme shyness. Despite his brilliance, he was a strange figure to many and stories of his social awkwardness pepper his biographies. Undoubtedly, he was withdrawn and introverted. It is said that he could talk to just one man at a time and only would communicate with women, even servants, through written notes. In fact, one story tells of an admirer heaping praise upon the reclusive scientist at a party. Cavendish became so uncomfortable that he ran from the room, out of the house, and into his carriage.

After his father's death, Henry became rich but rarely used his wealth. He went to the weekly Royal Society meeting dinners with exactly enough money, never less and never more. He also dressed in clothes that were considered old-fashioned for that time. However, he was quite charitable when worthy causes approached him.

The great chemist and inventor Humphry Davy, also a member of the Royal Society, left a descriptive view of Cavendish among his papers, published in *The Collected Works of Humphry Davy*. Davy's characterization is both candid and respectful: "Cavendish was a great man, with extraordinary singularities. His voice was squeaking, his manner nervous, he was afraid of strangers, and seemed, when embarrassed, even to articulate with difficulty."

Despite his eccentricities, Cavendish's fellow scientists held him in great esteem. His advice was often sought and accepted. It is thanks to these colleagues that the single drawing of Cavendish exists. He vehemently refused to sit for a portrait, and so artist William Alexander was invited to a Royal Society dinner. First, Alexander secretly sketched Cavendish's coat and hat, which he had left near the door. Then, during the meal, the artist drew Cavendish's profile as the shy man ate. Alexander later put the sketches together to create the drawing that today hangs in the British Museum in London.

Given Cavendish's shy reputation, his passion for science must have been overwhelmingly great. He was a faithful participant in his scientific organizations, especially the Royal Society. He performed a service to his fellow scientists by opening up the extensive library of his London house to them. Users could "check out" his books. Cavendish, too, recorded his name in the library's register.

Cavendish would likely be more celebrated if he had been in the race to publish his findings as others had been. But Henry Cavendish never sought fame or accolades, only knowledge. The great span and depth of his scientific contributions were only appreciated upon examination of his private papers after his death. However, many scientists today acknowledge that he was truly one of the most important scientists of the eighteenth century.

PERIODIC TABLE OF ELEMENTS

The periodic table organizes all the chemical elements into a simple chart according to the physical and chemical properties of their atoms. The elements are arranged by atomic number from 1 to 118. The atomic number is based on the number of protons in the nucleus of the atom. The atomic mass is the combined mass of protons and neutrons in the nucleus. Each element has a chemical symbol that is an abbreviation of its name. In some cases, such as potassium,

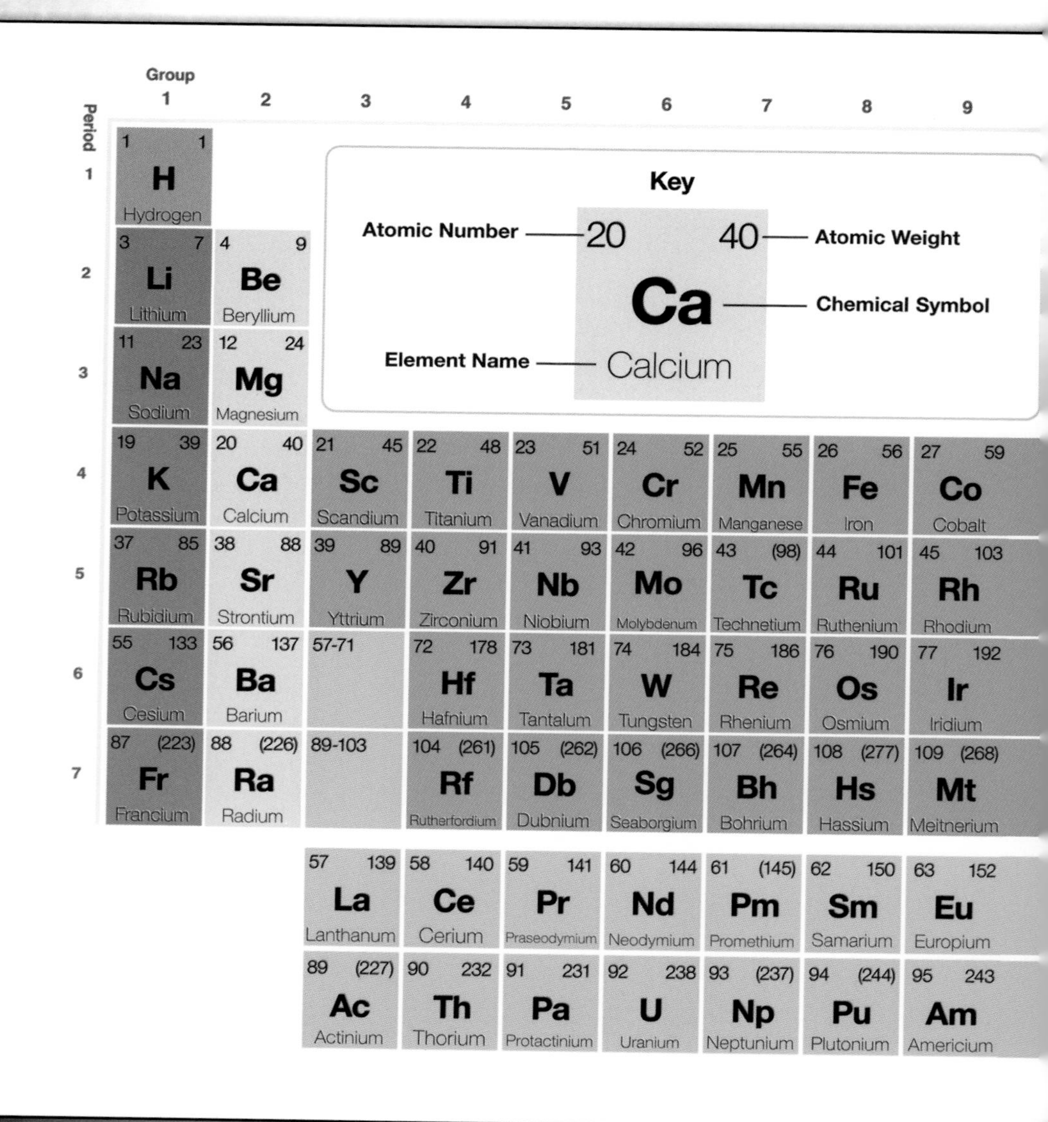

the symbol is an abbreviation of its Latin name ("K" stands for kalium). The name by which the element is commonly known is given in full underneath the symbol. The last item in the element box is the atomic mass. This is the average mass of an atom of the element.

Scientists have arranged the elements into vertical columns called groups and horizontal rows called periods. Elements in any one group all have the same number of electrons in their outer shell and have similar chemical properties. Periods represent the increasing number of electrons it takes to fill the inner and outer shells and become stable. When all the spaces have been filled (Group 18 atoms have all their shells filled), the next period begins.

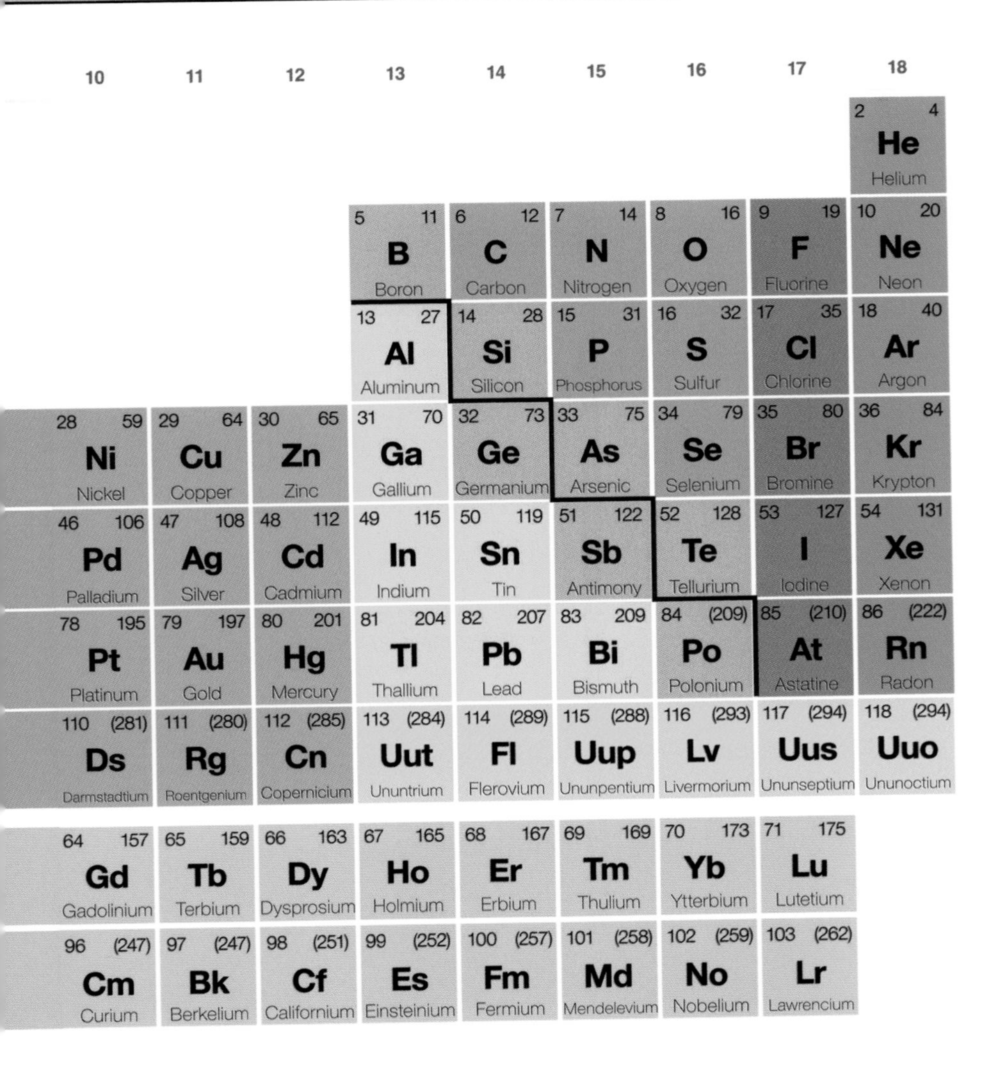

1731 Henry Cavendish is born October 10 in Nice, France.

1742 Cavendish begins Hackney Academy.

1749 Cavendish enters Peterhouse College, Cambridge.

1752 Cavendish finishes his schooling without a degree.

1758 Cavendish begins to attend meetings of the Royal Society of London with his father.

1760 Cavendish elected to Royal Society.

1766 Cavendish isolates hydrogen in a laboratory and wins the Copley Medal.

1771 Cavendish publishes a paper on electricity ("Attempt to Explain Some of the Principal Phenomena of Electricity by an Elastic Fluid").

1783 Cavendish's father dies, leaving him with substantial wealth.

1783 Cavendish is the first to produce water by combining hydrogen and oxygen but publishes his results after James Watt.

1784 Cavendish publishes a study on the composition of air ("Experiments on Air").

1785 Cavendish identifies another gas in common air, later discovered to be argon.

1798 Cavendish determines the density of Earth using a specially designed torsion balance.

1809 Cavendish's last paper, on improving astronomy tools, is published.

1810 Cavendish dies February 24 in London, England.

acid Substance that dissolves in water to form hydrogen ions (H+). Acids are neutralized by alkalis and have a pH below 7.

acid rain Rain that has an unusually high level of acidity resulting from pollution. Acid rain causes damage to forests, lakes, and rivers.

alkali Substance that dissolves in water to form hydroxide ions (OH-). Alkalis have a pH greater than 7 and will react with acids to form salts.

allotrope A different form of an element in which the atoms are arranged in a different structure.

amorphous Describes something that lacks a definite structure or shape.

atom The smallest independent building block of matter. All substances are made of atoms.

atomic mass The number of protons and neutrons in an atom's nucleus.

atomic number The number of protons in a nucleus.

boiling point The temperature at which a liquid turns into a gas.

bond A chemical connection between atoms.

catalyst Substance that speeds up a chemical reaction but is left unchanged at the end of the reaction.

chemical equation Symbols and numbers that show how reactants change into products during a chemical reaction.

chemical formula The letters and numbers that represent a chemical compound, such as "H_2O" for water.

chemical reaction The reaction of two or more chemicals (the reactants) to form new chemicals (the products).

chemical symbol The letters that represent a chemical, such as "Cl" for chlorine or "Na" for sodium.

combustion The reaction that causes burning. Combustion is generally a reaction with oxygen in the air.

compound Substance made from more than one element and that has undergone a chemical reaction.

condensation The change of state from a gas to a liquid.

conductor A substance that carries electricity and heat.

corrosion The slow wearing away of metals or solids by chemical attack.

covalent bond Bond in which atoms share one or more electrons.

crystal A solid made of regular repeating patterns of atoms.

crystal lattice The regular repeated structure found in crystalline solids.

density The mass of substance in a unit of volume.

deposit A mineral vein or ore inside another rock.

dissolve To form a solution.

distillation The process of evaporation and condensation used to separate a mixture of liquids according to their boiling points. Also a method of purifying a liquid.

electricity A stream of electrons or other charged particles moving through a substance.

electron A tiny, negatively charged particle that moves around the nucleus of an atom.

electronegativity The power of an atom to attract an electron. Nonmetals, which have only a few spaces in their outer shell, are the most electronegative. Metals, which have several empty spaces in their outer shell, tend to lose electrons in chemical reactions. Metals of this type are termed electropositive.

element A material that cannot be broken up into simpler ingredients. Elements contain only one type of atom.

energy The ability to do work.

energy level The shells around an atom each represent a different energy level. Those closest to the nucleus have the lowest energy.

fusion When small atoms fuse to make a single larger atom.

gas State in which particles are not joined and are free to move in any direction.

group A column of related elements in the periodic table.

halide A compound containing a halogen atom, such as chlorine or iodine.

heat The transfer of energy between atoms. Adding heat makes atoms move more quickly.

inorganic A compound that is not organic.

insoluble Describes a substance that cannot dissolve.

ion An atom that has lost or gained one or more electrons and gained an electric charge.

ionic bond Bond in which one atom gives one or more electrons to another atom.

isotope Atoms of a given element must have the same number of protons but can have different numbers of neutrons. These different versions of the same element are called isotopes.

liquid Substance in which particles are loosely bonded and are able to move freely around each other.

matter Anything that can be weighed.

melting point The temperature at which a solid changes into a liquid. When a liquid changes into a solid, this same temperature is called the freezing point.

molecule Two or more joined atoms that have a unique shape and size.

neutron One of the particles that make up the nucleus of an atom. Neutrons do not have any electric charge.

nitrogen fixation Processes that combines atmospheric nitrogen with other elements in a form that can be absorbed by plants.

noble gases A group of gases that rarely react with other elements.

nonmetal Any element that is not a metal. Most nonmetals are gases, such as hydrogen and argon.

nucleus The central part of an atom. The nucleus contains protons and neutrons. The exception is hydrogen, which contains only one proton.

organic A compound that is made of carbon and hydrogen.

oxide A compound that contains oxygen atoms.

oxidation The addition of oxygen to a compound. Also a reaction in which electrons are lost by an atom.

oxidation state A number used to describe how many electrons an atom can lose or gain.

oxidizer A substance that takes electrons from another atom or molecule to make its own outer shell stable.

ozone A form of oxygen in which three oxygen atoms join to form a molecule.

period A row of elements across the periodic table.

plasma "Fourth state of matter" in which atoms have lost some or all of their electrons.

pressure The force produced by pressing on something.

product The new substance or substances created by a chemical reaction.

proton A positively charged particle found in an atom's nucleus.

radiation The products of radioactivity—alpha and beta particles and gamma rays.

radioactive decay Process where small particles break off from an unstable nucleus.

reactants The ingredients necessary for a chemical reaction.

salt A compound made from positive and negative ions that forms when an alkali reacts with an acid.

semiconductor A substance that conducts heat and electricity, but only in certain circumstances.

shell The orbit of an electron. Each shell can contain a specific number of electrons and no more.

solid State of matter in which particles are held in a rigid arrangement.

solution A mixture of two or more elements or compounds in a single phase (solid, liquid, or gas).

solvent The liquid that dissolves a solute.standard conditions Normal room temperature and pressure.

state The form that matter takes—either a solid, a liquid, or a gas.

subatomic particles Particles that are smaller than an atom.

temperature A measure of how fast molecules are moving.

valence electrons The electrons in the outer shell of an atom.

volatile Describes a liquid that evaporates easily.

volume The space that a solid, liquid, or gas occupies.

American Association for the Advancement of Science (AAAS)
1200 New York, NY Avenue NW
Washington, DC 20005
(202) 326-6400
Web site: http://www.aaas.org
The American Association for the Advancement of Science is a nonprofit organization committed to advancing science around the world, in part through its publication *Science* as well as numerous newsletters.

American Hydrogen Association
P.O. Box 4205
Mesa, AZ 85201
(480) 234-5070
Web site: http://www.clean-air.org
The American Hydrogen Association is committed to achievements of prosperity without pollution and to close the information gap between researchers, industry and the public, drawing on worldwide developments concerning hydrogen, solar, wind, hydro, ocean and biomass resource materials, energy conversion, wealth-addition economics, and the environment.

Canadian Society for Chemistry
130 Slater Street, Suite 550
Ottawa, ON K1P 6E2
Canada
(888) 542-2242
Web site: http://www.cheminst.ca
The Canadian Society for Chemistry is a national, not-for-profit, professional association that unites chemistry students and professionals who work in industry, academia, and government.

Fuel Cell and Hydrogen Energy Association (FCHEA)
1211 Connecticut Avenue NW, Suite 600
Washington, DC 20036
(202) 261-1331
Web site: http://www.fchea.org
The Fuel Cell and Hydrogen Energy Association is dedicated to developing hydrogen energy technologies.

Los Alamos National Laboratory
P.O. Box 1663
Los Alamos, NM 87545
(505) 667-7000
Web site: http://www.lanl.gov
The mission of the Los Alamos National Laboratory is to "develop and apply science and technology to ensure the safety, security, and reliability of the U.S. nuclear deterrent; reduce global threats; and solve other emerging national security and energy challenges."

National Science Teachers Association (NSTA)
1840 Wilson Boulevard
Arlington VA 22201
(703) 243-7100
Web site: http://www.nsta.org
Founded in 1944, the National Science Teachers Association is committed to promoting excellence and

innovation in science teaching. Its sixty-thousand members include science teachers, science supervisors, administrators, scientists, business and industry representatives, and others involved in and committed to science education.

Society of Women Engineers
203 N. La Salle Street, Suite 1675
Chicago, IL 60601
(877) 793-4636
Web site: http://www.swe.org
Founded more than sixty years ago, the Society of Women Engineers gives women engineers a unique place and voice within the engineering industry. The organization is centered around a passion for members' success and continues to evolve with the challenges and opportunities reflected in today's exciting engineering and technology specialties.

WEB SITES

Due to the changing nature of Internet links, Rosen Publishing has developed an online list of Web sites related to the subject of this book. This site is updated regularly. Please use this link to access the list:

http://www.rosenlinks.com/CORE/Nonme

Atkins, P. W. *Reactions: The Private Life of Atoms*. Oxford, England: Oxford University Press, 2011.

Cobb, C., and Fetterolf, M. L. *The Joy of Chemistry: The Amazing Science of Familiar Things*. Amherst, NY: Prometheus Books, 2005.

Gray, Theodore W., and Nick Mann. *The Elements: A Visual Exploration of Every Known Atom in the Universe*. New York, NY: Black Dog & Leventhal, 2012.

Gupta, Ram B. *Hydrogen Fuel: Production, Transport, and Storage*. Boca Raton, FL: CRC, 2009.

Halka, Monica, and Brian Nordstrom. *Halogens and Noble Gases*. New York, NY: Facts on File, 2010.

Halka, Monica, and Brian Nordstrom. *Nonmetals*. New York, NY: Facts on File, 2010.

Hoffmann, Peter, and Byron L. Dorgan. *Tomorrow's Energy: Hydrogen, Fuel Cells, and the Prospects for a Cleaner Planet*. Cambridge, MA: MIT, 2012.

Holland, Geoffrey B., and James J. Provenzano. *The Hydrogen Age: Empowering a Clean-Energy Future*. Salt Lake City, UT: Gibbs Smith, 2007.

Kean, Sam. *The Disappearing Spoon: And Other True Tales of Madness, Love, and the History of the World from the Periodic Table of the Elements*. New York, NY: Back Bay, 2011.

La, Bella Laura. *The Oxygen Elements: Oxygen, Sulfur, Selenium, Tellurium, Polonium*. New York, NY: Rosen Publishing, 2010.

O'Hayre, Ryan P. *Fuel Cell Fundamentals*. Hoboken, NJ: John Wiley & Sons, 2009.

Roston, Eric. *The Carbon Age: How Life's Core Element Has Become Civilization's Greatest Threat*. New York, NY: Walker & Co., 2009.

Roza, Greg. *The Halogen Elements: Fluorine, Chlorine, Bromine, Iodine, Astatine*. New York, NY: Rosen Publishing, 2010.

Sherman, Josepha. *Henry Cavendish and the Discovery of Hydrogen*. Hockessin, DE: Mitchell Lane, 2005.

Yorifuji, Bunpei. *Wonderful Life with the Elements: An Adventure Through the Periodic Table*. San Francisco, CA: No Starch, 2012.

INDEX

A

B

C

I

K

L

M

N

O

P

R

S

T

PHOTO CREDITS

Cover, p. 3, interior pages (borders) Melissa Carroll/E+/Getty Images; p. 6 Scowen/Time & Life Pictures/Getty Images; pp. 7 (top), 17 AFP/Getty Images; pp. 8, 63 Keystone-France/Gamma-Keystone/Getty Images; pp. 9 (left), 12, 20, 22, 23, 28, 30 (bottom), 36, 38, 45, 47, 50, 54, 55 (top), 58 (left), 68 iStockphoto.com/Thinkstock; pp. 10, 29 (left), 48, 76 Charles D. Winters/Photo Researchers/Getty Images; p. 11 Science Source; p. 14 New York Daily News/Getty Images; p. 15 (top) Huriye Akinci Iriyari/E+/Getty Images; p. 16 Jupiterimages/Photos.com/Thinkstock; p. 17 AFP/Getty Images; p. 24 Bloomberg/Getty Images; p. 25 David Cannon/Getty Images; p. 26 Hemera Technologies/AbleStock.com/Thinkstock; p. 31 (top) Photos.com/Thinkstock; p. 31 (bottom) Universal Images Group/Getty Images; p. 32 Digital Vision/Thinkstock; p. 33 Filip Singer/EPA/Landov; 34 (top) Thomas Northcut/Digital Vision/Thinkstock; 34 (bottom) W. K. Fletcher/Photo Researchers/Getty Images; p. 35 Charles D Winters/Science Source; p. 39 Visuals Unlimited, Inc./GIPhotoStock/Getty Images; p. 40 John Foxx/Stockbyte/Thinkstock; p. 41 (bottom) Colin Milkins Collection/Oxford Scientific/Getty Images; p. 42 David Hay Jones/Science Source; p. 43 (bottom) Hulton Archive/Getty Images; p. 44 Belinda Images/SuperStock; p. 46 Stocktrek Images/Richard Roscoe/Getty Images; p. 49 Hemera/Thinkstock; pp. 52, 77 Science Source/Photo Researchers/Getty Images; p. 55 (bottom) Andrew Lambert Photography/Science Source; p. 56 (left) © iStockphoto.com/crispphotography; p. 56 (right) © iStockphoto.com/farelka; p. 58 (right) Mondadori/Getty Images; p. 59 (top) Sebastian Radu/Shutterstock.com; p. 59 (bottom) kataijudit/Shutterstock.com; p. 60 F. Jimenez Meca/Shutterstock.com; p. 61 Keith Brofsky/Photodisc/Getty Images; p. 62 (top) Dr. Kenneth Greer/Visuals Unlimited/Getty Images; p. 62 (bottom) Ryan McVay/Digital Vision/Thinkstock; pp. 64, 69, 75 Science & Society Picture Library/Getty Images; p. 66 NASA; p. 70 Nick Poling/Shutterstock.com; p. 71 Alexander Tsiaras/Science Source; p. 72 Top Photo Group/Thinkstock; p. 74 Photo Researchers/Getty Images; p. 79 © SSPL/The Image Works; pp. 82–83 Kris Everson; top of pp. 6, 12, 28, 40, 54, 66, 74 © iStockphoto.com/aleksandar velasevic; all other photos and illustrations © Brown Bear Books Ltd.